Advances in 3OM: Opto-Mechatronics, Opto-Mechanics, and Optical Metrology

Advances in 3OM: Opto-Mechatronics, Opto-Mechanics, and Optical Metrology

Guest Editors

Virgil-Florin Duma
Guillermo Garcia-Torales
Tomohiko Hayakawa

MDPI

Basel • Beijing • Wuhan • Barcelona • Belgrade • Novi Sad • Cluj • Manchester

Guest Editors
Virgil-Florin Duma
Measurements and Optical Electronics
Polytechnic University of Timisoara
Timisoara
Romania

Guillermo Garcia-Torales
Departamento de Electro-Fotónica
Universidad de Guadalajara
Guadalajara
Mexico

Tomohiko Hayakawa
Research Institute for Science and Technology
Tokyo University of Science
Tokio
Japan

Editorial Office
MDPI AG
Grosspeteranlage 5
4052 Basel, Switzerland

This is a reprint of the Special Issue, published open access by the journal *Photonics* (ISSN 2304-6732), freely accessible at: www.mdpi.com/journal/photonics/special_issues/3OM.

For citation purposes, cite each article independently as indicated on the article page online and as indicated below:

Lastname, A.A.; Lastname, B.B. Article Title. *Journal Name* **Year**, *Volume Number*, Page Range.

ISBN 978-3-7258-4536-1 (Hbk)
ISBN 978-3-7258-4535-4 (PDF)
https://doi.org/10.3390/books978-3-7258-4535-4

Contents

About the Editors

Virgil-Florin Duma

Virgil-Florin Duma is a Professor at the Polytechnic University of Timisoara (UPT). In 2008, he proposed the idea of the 3OM field (in Opto-Mechatronics, Opto-Mechanics and Opt. Metrology) and founded the 3OM Group. He received his PhD cum laude from UPT in 2001, where he was admitted and graduated valedictorian in Fine Mechanics and Optics in 1986 and 1991, respectively. In 2009-2010, he was awarded a Fulbright Senior Research Fellowship at The Institute of Optics, University of Rochester, NY, USA, where he is a Visiting Researcher. He is an Honorary Professor at the University of Kent, UK, and an Adjunct Professor at several universities, including at the Polytechnic University of Bucharest and at the Victor Babes University of Medicine and Pharmacy of Timisoara. He is the author of >70 papers in WOS journals (with a total IF>150), >130 conference papers in WOS, and 15 books, and he has been awarded three patents. Prof. Duma secured funds in excess of EUR 2.5 M for his group, from more than 10 research grants. He has supervised >15 PhD students and numerous undergraduate students. He has delivered >50 invited lectures at int'l univ. and conferences. Prof. Duma has been a reviewer for >40 WOS journals and the chair of several SPIE-affiliated conferences, including Advances in 3OM. He is a member of OSA, senior member of SPIE, and life member of the Fulbright Association.

Guillermo Garcia-Torales

Dr. Guillermo García Torales is a Full Professor at the University of Guadalajara (CUCEI), Mexico, where he has served since 2001. He earned his Ph.D. in Optics in 2001 from the Centro de Investigaciones en Óptica (CIO), with a dissertation titled "Vectorial Displacement Interferometer", under the supervision of Dr. Marija Strojnik. He completed a postdoctoral stay at the International Center for Water Resources (CIRA) in Toluca, where he developed particle image velocimetry (PIV) systems for fluid analysis. He founded the Laboratory of Optical and Electronic Metrology and Instrumentation at the University of Guadalajara, where his research focuses on optical interferometry, electro-optical instrumentation, and optical metrology, including the design of non-invasive measurement systems. He has authored over 100 scientific publications—40 of which appear in JCR-indexed journals—and has supervised more than 50 graduate theses. Dr. García Torales currently holds a Level II distinction in Mexico's National System of Researchers (SNI) and is a member of the Mexican Academy of Sciences, the Mexican Academy of Optics, and the Mexican Physical Society. He is also a Senior Member of SPIE, advisor to the SPIE and OPTICA student chapters, and a member of the SPIE Conference Program Committee. He has also served as a Guest Editor for the journal *Photonics*.

Tomohiko Hayakawa

Tomohiko Hayakawa is an Associate Professor at the Research Institute of Science and Technology, Tokyo University of Science. He earned his Ph.D. in Information Science and Technology from the University of Tokyo in 2016, following positions as a specially appointed project assistant professor and assistant professor at the Graduate School of Information Science and Technology. In 2020, he joined the University of Tokyo Information Technology Center as a project associate professor and stayed until 2024 (continued in FY2023 in a concurrent position). Dr. Hayakawa's research focuses on high-speed vision systems, dynamic vision, and active perception. As the group leader of the Active Perception group, he has developed innovative imaging techniques, such as motion-blur compensated visual inspection for high-speed travel in tunnel infrastructure. His work has been recognized with numerous awards, including the 2024 IEICE Communication Society Magazine Paper Award, the 2021 Robot Award (Research & Development Division), and the 2020 VR Society Paper Award. He has led several major research projects, including serving as the principal investigator on a Grant in Aid for Scientific Research (B) on dynamic phosphorescent markers for label-free tracking, and is currently the PI for a rotating-prism-based high-speed imaging project. His academic contributions span top-tier journals such as *Applied Optics*, *Optics Express*, *Sensors*, *PLoS ONE*, and *Journal of Robotics and Mechatronics*, as well as numerous invited and international conference presentations. Dr. Hayakawa is also an active member of professional societies including VR Society of Japan, the Society of Instrument and Control Engineers, and the Japan Society of Civil Engineers. His ongoing research aims to bridge the gap between perception science and applied infrastructure monitoring through high-speed optical sensing technologies, contributing to real-time, non-invasive diagnostic systems.

photonics

MDPI

Editorial

Introduction to Special Issue on "Advances in 3OM: Opto-Mechatronics, Opto-Mechanics, and Optical Metrology"

Virgil-Florin Duma [1,2,3,*], Guillermo Garcia-Torales [4] and Tomohiko Hayakawa [5]

1 3OM Optomechatronics Group, Department of Measurements and Optical Electronics, Faculty of Electronics, Telecommunications, and Information Technology, Polytechnic University of Timisoara, 2 Vasile Parvan Ave., 300223 Timisoara, Romania
2 Faculty of Engineering, "Aurel Vlaicu" University of Arad, 2 Elena Dragoi Street, 310177 Arad, Romania
3 Center of Research and Development for Mechatronics, National University of Science and Technology POLITEHNICA Bucharest, 060042 Bucharest, Romania
4 Departamento de Electro-Fotónica, Centro Universitario de Ciencias Exactas e Ingenierías (CUCEI), Universidad de Guadalajara (U. de G.), Blvd. M. García Barragán 1421, Guadalajara 44410, Jalisco, Mexico; garcia.torales@academicos.udg.mx
5 Research Institute for Science and Technology, Tokyo University of Science, Shinjuku-ku, Tokyo 162-8601, Japan; hayakawa@ishikawa-vision.org
* Correspondence: virgil.duma@upt.ro; Tel.: +40-751-511451

1. Introduction

The 3OM concept was introduced in 2008 and combines three complementary domains: opto-mechatronics, opto-mechanics, and optical metrology.

Opto-Mechatronics was introduced in 2005 by Prof. Hyungsuck Cho in a seminal book with the same name [1]. It combines optics and mechatronics, the latter of which is a blend of precision mechanics, electronics, control, and automation, as well as IT. This multidisciplinary approach is beneficial in the engineering and design of complex systems, adding value to each of the component fields. A good example can be seen in the form of laser scanning techniques [2,3], which involve, for example, galvanometer-based [4–6], polygon mirror [7–9], or Risley prisms [10–13] and can be designed for a wide range of applications, from commercial to industrial and high-end purposes. These include applications in areas as diverse as biomedical imaging [14] and remote sensing [15].

Opto-Mechanics addresses positioning issues and errors in optical components [16,17] and tolerances of both mechanical and optical parts, as well as methods for tackling these aspects. Researchers in this field strive to fill the gap between the high (theoretical) precision and requirements of optical design and the practical capabilities that are inherent in mechanical technologies [2]. Another aspect that is often overlooked is related to the kinematics and dynamics of systems with optical moving parts (i.e., mirrors, lenses, and prisms). This imposes finite element analyses (FEAs), especially of fast-moving (for example, rotational) parts, in order to control structural integrity issues, as well as deformation levels [3,9]. FEAs and various other mechanical analyses are necessary, even for systems with non-moving or slow-moving (but large and heavy) optical parts (and mounts), for example, with refractive elements (e.g., Risley prisms [18]) or segmented mirrors [19]. These aspects bridge opto-mechanics and opto-mechatronics. This relationship is strengthened by the necessity for control and automation (as included in mechatronics) in (precise positioning and displacements in) systems with moving components [20–22].

Optical Metrology comprises a wide area of applications that are related to both domains above, especially opto-mechatronics. It includes domains that range from industrial measurements to imaging, with the latter being used both in industry, with non-destructive

https://doi.org/10.3390/photonics12060557

testing (NDT), and in the biomedical field, with a wide range of techniques, for example, optical coherence tomography (OCT) [23–25], confocal microscopy [26], or photoacoustics.

3OM also enhances laser manufacturing, which, in turn, implies robotics, machine vision, and machine learning, the latter with AI algorithms. Besides biomedical imaging, it is closely connected to various laser techniques in medicine. Finally, its field of interest extends to include remote sensing, as well as security and defense.

Regarding devices and specific technologies, we must highlight the rapidly evolving field of micro-electro-mechanical systems (MEMSs), which has applications in all the above fields [24,27–30].

2. An Overview of the Articles in This Special Issue

In the work described in Contribution 1, the reflectivity of a cholesteric liquid crystal was numerically investigated for an anisotropic defect layer inside. The optical phenomena were modeled for different parameters of the system, as well as adjusting the external applied electric field. The possibilities of controlling the reflection spectrum were obtained.

The authors of Contribution 2 studied a shock wave that may be induced with a high-power laser pulse. A theoretical model of laser material processing was developed, with experimental validation via piezo-resistive methods. The field of laser micromachining was also approached in Contribution 3 for laser drilling. Optical detection achieved a >95% accuracy for tens-of-micrometer-deep holes with a micrometer diameter. Contribution 4 focuses on optical vortices (i.e., ultra-intense laser pulses with helical phases), assessing their field distributions.

The authors of Contribution 5 approached an opto-mechatronic device, introducing and demonstrating a low-cost 3D-printed 1-DOF laser scanner. Its parameters were determined and validated experimentally, offering an alternative to the more expensive (and most common) galvanometer scanners.

Optical metrology was targeted for Contribution 6, with the Digital Image Correlation (DIC) method used for displacement- and stress–strain-invariant deformation measurements. A reverse retrieval strategy was also developed. The DIC measurements allowed for robust and efficient displacement invariance measurement, with an average accuracy of 0.1% observed for the stress–strain results.

A system development is proposed in Contribution 7, with a broadband mode coupler for multimode OCT in the O-band (1.26–1.36 μm). Key design parameters were studied. Reflected signals from the sample were separated using the same fiber device before interfering with the reference light, which had not been previously possible. The proposed fiber device is expected to represent a key component in efficiently achieving multimode OCT operation with better signal collection efficiency and improved penetration depth for deep tissue imaging.

Contribution 8 explores high-performance focal plane arrays and their fields of application, including remote sensing, astronomical, and surveillance instruments. Whereas in the analysis of an instrument performance analysis, it is assumed that the image of a point source is at the center of a detector pixel, in reality, it may fall at any position in the detector pixel. Pointing errors and jitter may lead to errors of up to 20%. Critical factors that impact the performance estimate include the optical centroid efficiency (OCE) and the ensquared energy (i.e., the energy on the rectangular detector pixel (EOD)). Simulations were performed for imaging with and without a generalized rectangular central obscuration. Contribution 10 represents the second part of this study, analyzing the performance of the OCE vs. the EOD. The three Seidel primary aberrations of an optical component (i.e., spherical, coma, and astigmatism, plus defocus) were considered. The study concluded that the choice of the larger pixel might be advantageous for low-aberration instruments in dynamic and unpredictable environments. Thus, for pixels larger than a certain threshold, a small pixel shows better performance in the face of jitter, misalignment, and other environmental conditions.

The authors of Contribution 9 utilized computer simulations to approach microstructure observations using speckle interferometry. This study demonstrates that the separation of two close points is not impossible when coherent light is used. The condition examined comprised different light phases between the two points. This discussion on the resolution of microstructure observations based on speckle interferometry led to a new interpretation of the Rayleigh criterion in super-resolution techniques.

3. Conclusions

In summary, 3OM research is multidisciplinary, bridging several optical, mechanical, and electrical engineering fields. The ten contributions to this Special Issue represent an enticing taster of the many areas of interest involving 3OM. Two of the works were prepared for the second edition of the International Conference Advances in 3OM [31]. During the publishing process, the third edition of this Conference was under development [32], and, linked to it, the second edition of this Special Issue of *Photonics* was launched [33]. Another Special Issue, this time of *Sensors*, also addresses a 3OM topic of high interest: laser scanning and its various applications [34]. We hope that these Special Issues offer readers an insight into these fields and inspire scholars to contribute to advancing our knowledge on these topics for the benefit of humankind.

Funding: V.-F. Duma was funded during the development of this Special Issue by the Romanian Ministry of Research, Innovation, and Digitization, CNCS/CCCDI–UEFISCDI, projects PN-III-P4-ID-PCE-2020-2600 and PN-III-P2-2.1-PED-2020-4423, both within PNCDI III (http://3om-group-optomechatronics.ro/, accessed on 1 April 2025). This research is currently supported by the Romanian IPCEI (Important Project of Common European Interest) on microelectronics, via Polytechnic University of Timisoara and Continental Automotive Romania, as well as by the European Union through COST Action CA21159 (PhoBioS).

Acknowledgments: The Guest Editors of this Special Issue, "Advances in 3OM: Opto-Mechatronics, Opto-Mechanics, and Optical Metrology", would like to express our sincere thanks and deep appreciation to all authors published in this Special Issue for their contribution to its success. We also thank our reviewers, as well as the *Photonics* editors and staff for their outstanding support.

Conflicts of Interest: The authors declare no conflicts of interest. The funders had no role in the design of the study; in the collection, analyses, or interpretation of data; in the writing of the manuscript; or in the decision to publish the results.

List of Contributions:

1. Grzelczyk, D.; Awrejcewicz, J. Reflectivity of Cholesteric Liquid Crystals with an Anisotropic Defect Layer Inside. *Photonics* **2020**, *7*, 58. https://doi.org/10.3390/photonics7030058.
2. Gonzalez-Romero, R.; Strojnik, M.; Garcia-Torales, G.; Gomez-Rosas, G. Frequency Dependence of a Piezo-Resistive Method for Pressure Measurements of Laser-Induced Shock Waves in Solids. *Photonics* **2021**, *8*, 120. https://doi.org/10.3390/photonics8040120.
3. Chen, X.; Xu, Y.; Chen, N.-K.; Shy, S.; Chui, H.-C. In Situ Depth Measurement of Laser Micromachining. *Photonics* **2021**, *8*, 493. https://doi.org/10.3390/photonics8110493.
4. Talposi, A.-M.; Iancu, V.; Ursescu, D. Influence of Spatio-Temporal Couplings on Focused Optical Vortices. *Photonics* **2022**, *9*, 389. https://doi.org/10.3390/photonics9060389.
5. Shen, C.-K.; Huang, Y.-N.; Liu, G.-Y.; Tsui, W.-A.; Cheng, Y.-W.; Yeh, P.-H.; Tsai, J.-c. Low-Cost 3D-Printed Electromagnetically Driven Large-Area 1-DOF Optical Scanners. *Photonics* **2022**, *9*, 484. https://doi.org/10.3390/photonics9070484.

6. Jain, A.; Mishra, A.; Tiwari, V.; Singh, G.; Singh, R.P.; Singh, S. Deformation Measurement of a SS304 Stainless Steel Sheet Using Digital Image Correlation Method. *Photonics* **2022**, *9*, 912. https://doi.org/10.3390/photonics9120912.
7. Hu, D.J.J.; Liu, L.; Dong, H.; Zhang, H. Design of a Broadband Fiber Optic Mode Coupler for Multimode Optical Coherence Tomography. *Photonics* **2023**, *10*, 162. https://doi.org/10.3390/photonics10020162.
8. Strojnik, M.; Bravo-Medina, B.; Martin, R.; Wang, Y. Ensquared Energy and Optical Centroid Efficiency in Optical Sensors: Part 1, Theory. *Photonics* **2023**, *10*, 254. https://doi.org/10.3390/photonics10030254.
9. Arai, Y.; Chen, T. Simulation-Based Considerations on the Rayleigh Criterion in Super-Resolution Techniques Based on Speckle Interferometry. *Photonics* **2023**, *10*, 374. https://doi.org/10.3390/photonics10040374.
10. Strojnik, M.; Martin, R.; Wang, Y. Ensquared Energy and Optical Centroid Efficiency in Optical Sensors: Part 2, Primary Aberrations. *Photonics* **2024**, *11*, 855. https://doi.org/10.3390/photonics11090855.

References

1. Cho, H. *Optomechatronics: Fusion of Optical and Mechatronic Engineering*, 1st ed.; CRC Press: Boca Raton, FL, USA, 2005. [CrossRef]
2. Beiser, L.; Johnson, B. Scanners. In *Handbook of Optics*, 3rd ed.; Bass, M., Ed.; McGraw Hill Inc.: New York, NY, USA, 2009; pp. 30.1–30.68.
3. Marshall, G.F.; Stutz, G.E. (Eds.) *Handbook of Optical and Laser Scanning*, 2nd ed.; CRC Press: London, UK, 2011.
4. Li, Y. Single-mirror beam steering system: Analysis and synthesis of high-order conic-section scan patterns. *Appl. Opt.* **2008**, *47*, 386–398. [CrossRef] [PubMed]
5. Montagu, J. Scanners-galvanometric and resonant. In *Encyclopedia of Optical and Photonic Engineering*; CRC Press: Boca Raton, FL, USA, 2003; pp. 2465–2487.
6. Duma, V.-F. Laser scanners with oscillatory elements: Design and optimization of 1D and 2D scanning functions. *Appl. Math. Model.* **2019**, *67*, 456–476. [CrossRef]
7. Li, Y.; Katz, J. Asymmetric distribution of the scanned field of a rotating reflective polygon. *Appl. Opt.* **1997**, *36*, 342–352. [CrossRef] [PubMed]
8. Oh, W.Y.; Yun, S.H.; Tearney, G.J.; Bouma, B.E. 115 kHz tuning repetition rate ultrahigh-speed wavelength-swept semiconductor laser. *Opt. Lett.* **2005**, *30*, 3159–3161. [CrossRef]
9. Duma, V.-F.; Duma, M.-A. Optomechanical Analysis and Design of Polygon Mirror-Based Laser Scanners. *Appl. Sci.* **2022**, *12*, 5592. [CrossRef]
10. Yang, Y. Analytic solution of free space optical beam steering using Risley prisms. *J. Light. Technol.* **2008**, *26*, 3576–3583. [CrossRef]
11. Zhou, Y.; Fan, D.; Fan, S.; Chen, Y.; Liu, G. Laser scanning by rotating polarization gratings. *Appl. Opt.* **2016**, *55*, 5149–5157. [CrossRef]
12. Li, A.; Yi, W.; Zuo, Q.; Sun, W. Performance characterization of scanning beam steered by tilting double prisms. *Opt. Express* **2016**, *24*, 23543–23556. [CrossRef]
13. Duma, V.-F.; Dimb, A.-L. Exact Scan Patterns of Rotational Risley Prisms Obtained with a Graphical Method: Multi-Parameter Analysis and Design. *Appl. Sci.* **2021**, *11*, 8451. [CrossRef]
14. Leung, M.K.K.; Mariampillai, A.; Standish, B.A.; Lee, K.K.C.; Munce, N.R.; Vitkin, I.A.; Yang, V.X.D. High-power wavelength-swept laser in Littman telescope-less polygon filter and dual-amplifier configuration for multichannel optical coherence tomography. *Opt. Lett.* **2009**, *34*, 2814–2816. [CrossRef]
15. Li, A.; Liu, X.; Sun, J.; Lu, Z. Risley-prism-based multi-beam scanning LiDAR for high-resolution three-dimensional imaging. *Opt. Lasers Eng.* **2022**, *150*, 106836. [CrossRef]
16. Bravo-Medina, B.; Strojnik, M.; Garcia-Torales, G.; Torres-Ortega, H.; Estrada-Marmolejo, R.; Beltrán-González, A.; Flores, J.L. Error compensation in a pointing system based on Risley prisms. *Appl. Opt.* **2017**, *56*, 2209–2216. [CrossRef] [PubMed]
17. Ge, Y.; Liu, J.; Xue, F.; Guan, E.; Yan, W.; Zhao, Y. Effect of mechanical error on dual-wedge laser scanning system and error correction. *Appl. Opt.* **2018**, *57*, 6047–6054. [CrossRef]
18. Li, A.; Jiang, X.; Sun, J.; Bian, Y.; Wang, L.; Liu, L. Radial support analysis for large-aperture rotating wedge prism. *Opt. Laser Technol.* **2012**, *44*, 1881–1888. [CrossRef]
19. Zhao, K.; Song, L.; Sun, D.; Huang, Q.; Guo, R.; Tian, G.; He, J. Design and Research of Large-Scale Curvature Adjustment Optical Mirror. *Photonics* **2025**, *12*, 338. [CrossRef]

20. Yoo, H.W.; Ito, S.; Schitter, G. High speed laser scanning microscopy by iterative learning control of a galvanometer scanner. *Control. Eng. Pract.* **2016**, *50*, 12–21. [CrossRef]
21. Hayakawa, T.; Watanabe, T.; Senoo, T.; Ishikawa, M. Gain-compensated sinusoidal scanning of a galvanometer mirror in proportional-integral-differential control using the pre-emphasis technique for motion-blur compensation. *Appl. Opt.* **2016**, *55*, 5640–5646. [CrossRef]
22. Xiao, H.; Jiang, T.; Chen, X. Rejecting fast narrow-band disturbances with slow sensor feedback for quality beam steering in selective laser sintering. *Mechatronics* **2018**, *56*, 166–174. [CrossRef]
23. Drexler, W.; Liu, M.; Kumar, A.; Kamali, T.; Unterhuber, A.; Leitgeb, R.A. Optical coherence tomography today: Speed, contrast, and multimodality. *J. Biomed. Opt.* **2014**, *19*, 071412. [CrossRef]
24. Lu, C.D.; Kraus, M.F.; Potsaid, B.; Liu, J.J.; Choi, W.; Jayaraman, V.; Cable, A.E.; Hornegger, J.; Duker, J.S.; Fujimoto, J.G. Handheld ultrahigh speed swept source optical coherence tomography instrument using a MEMS scanning mirror. *Biomed. Opt. Express* **2014**, *5*, 293–311. [CrossRef]
25. Monroy, G.L.; Won, J.; Spillman, D.R.; Dsouza, R.; Boppart, S.A. Clinical translation of handheld optical coherence tomography: Practical considerations and recent advancements. *J. Biomed. Opt.* **2017**, *22*, 121715. [CrossRef] [PubMed]
26. Warger, W.C., II; DiMarzio, C.A. Dual-wedge scanning confocal reflectance microscope. *Opt. Lett.* **2007**, *32*, 2140–2142. [CrossRef] [PubMed]
27. Mu, X.; Zhou, G.; Yu, H.; Du, Y.; Feng, H.; Tsai, J.M.L.; Chau, F.S. Compact MEMS-driven pyramidal polygon reflector for circumferential scanned endoscopic imaging probe. *Opt. Express* **2012**, *20*, 6325–6339. [CrossRef] [PubMed]
28. Strathman, M.; Liu, Y.; Keeler, E.G.; Song, M.; Baran, U.; Xi, J.; Sun, M.-T.; Wang, R.; Li, X.; Lin, L.Y. MEMS scanning micromirror for optical coherence tomography. *Biomed. Opt. Express* **2015**, *6*, 211. [CrossRef]
29. Pan, T.; Gao, X.; Yang, H.; Cao, Y.; Zhao, H.; Chen, Q.; Xi, H. A MEMS Mirror-Based Confocal Laser Endomicroscope with Image Distortion Correction. *IEEE Photonics J.* **2023**, *15*, 3900408. [CrossRef]
30. Gora, M.J.; Suter, M.J.; Tearney, G.J.; Li, X. Endoscopic optical coherence tomography: Technologies and clinical applications [Invited]. *Biomed. Opt. Express* **2017**, *8*, 2405–2444. [CrossRef]
31. 2nd International Conference 'Advances in 3OM' 2023. Available online: https://3om-group-optomechatronics.ro/advances-in-3om-conference-2023/ (accessed on 17 May 2025).
32. 3OM Optomechatronics Group. Available online: https://3om-group-optomechatronics.ro/home/ (accessed on 17 May 2025).
33. 'Advances in 3OM: Opto-Mechatronics, Opto-Mechanics and Optical Metrology' Special Issue in Photonics, 2nd ed. Available online: https://www.mdpi.com/journal/photonics/special_issues/NS5DPR2QS9 (accessed on 17 May 2025).
34. 'Laser Scanning and Applications' Special Issue in Sensors. Available online: https://www.mdpi.com/journal/sensors/special_issues/1W05Q2ZEC8 (accessed on 17 May 2025).

photonics

MDPI

Article

Reflectivity of Cholesteric Liquid Crystals with an Anisotropic Defect Layer Inside

Dariusz Grzelczyk * and Jan Awrejcewicz

Department of Automation, Biomechanics and Mechatronics, Lodz University of Technology, 90-924 Lodz, Poland; jan.awrejcewicz@p.lodz.pl

* Correspondence: dariusz.grzelczyk@p.lodz.pl

Received: 15 June 2020; Accepted: 7 August 2020; Published: 10 August 2020

Abstract: In this study, first, we numerically investigated the reflectivity of a cholesteric liquid crystal with an anisotropic defect layer inside. To model optical phenomena in the examined system, a 4×4 matrix method was employed. The tests were carried out for different thicknesses of the whole system, different thicknesses of the defect layer, as well as different defect layer locations inside the cell. Next, a cholesteric liquid crystal comprising a defect layer and held between two parallel electrical conductors was also considered. In this case, the optical properties of the system could also be adjusted by an external applied electric field. Some interesting simulation results of the reflection coefficient (i.e., the fraction of electromagnetic energy reflected) were obtained, illustrated, and discussed. The simulation results showed a significant influence of both the defect and the external electric field on the selective reflection phenomenon, and the possibility of controlling the shape of the reflection spectrum. Finally, some potential applications of the analyzed optical system were discussed.

Keywords: photonic materials; cholesteric liquid crystal; optical metrology; reflectivity; reflection; bandgap; defect mode; defect layer; director field; electric field

---

1. Introduction

Photonic crystals consisting of artificial or organic self-organizing periodic structures have opened up new possible applications in modern electronic devices. Some of the possible practical applications of such materials in different branches of science and technology can be found, for instance, in recent review papers [1–5]. Especially attractive is the use of cholesteric liquid crystals (CLCs) since the macroscopic properties of such optical structures can be manipulated by external stimuli.

CLCs are organic materials with unique and useful optical properties. They are composed of optically anisotropic and uniaxial elements (molecules), with one major and two minor principal axes, which are also mutually perpendicular locally. In cholesteric liquid crystals, there is a long-range orientational order of the elongated molecules with locally preferred common direction called the director. Whereas in nematic liquid crystals, the elastic forces and torques tend to establish a uniform orientation of the director, the chiral molecules in CLCs lead in equilibrium to a helical arrangement with the director perpendicular to the helix axis [6]. This way, they have a one-dimensional periodic modulation of dielectric constants, which leads to interesting optical effects.

Owing to the optical anisotropy and helical twist, CLCs are characterized by the so-called selective reflection. It means that for the light incident along the helical axis (normal incidence), the circularly polarized light with the same handedness as that of CLC, is totally reflected in the specific wavelength range. The edges of this bandwidth are defined by two sharp optical modes which correspond to circularly polarized standing waves in which the electric field in the bulk of the sample is aligned along either the ordinary or the extraordinary molecular axes [7]. This unique feature of the selective reflection is widely used in numerous applications such as reflective displays, tunable color filters, or mirrorless lasers [8].

In isotropic periodic media, defects have been created by the removal or addition of a material, or by an alteration of the refractive index of one or a number of elements of photonic crystals [9]. In the anisotropic optical structures considered in this study, defects can be introduced by replacing a part of the host medium with a material that has a different dielectric constant and can make defect modes in the forbidden bands [10]. The defect makes such structures more useful (such as doped semiconductors) and can be successfully used for constructing narrowband filters [11–13], low threshold lasers [14,15], and other optical devices. A broad overview of the properties of all currently existing defective modes in CLCs, including both the proposed theoretical approaches and the performed experimental observations, can be found in monography [16]. In particular, efficient applications of localized modes for low-threshold lasing at the frequencies of localized modes are presented in this study as the most recent achievements in this area.

Generally, two ways to introduce a defect in CLCs have been proposed (and discussed in more detail below): either by replacing a thin layer of the CLC with an isotropic material [10], or by introducing a phase jump in the cholesteric helix [7]. In the first case, in analogy with isotropic periodic structures, a defect can be produced in a helical structure by adding an isotropic layer in the middle of a CLC. In the second case, a chiral twist defect can be created by rotating one part of the sample about its helical axis without separating the two parts.

Yang et al. [10] theoretically studied the reflection and transmission of light from a one-dimensional system consisting of two layers of cholesteric liquid crystal sandwiching a thin layer of the isotropic medium as a defect. Reflection spectra for normal incidence, calculated by using the so-called "faster Berreman method", were studied to obtain the wavelengths and intensities of defect modes when the refractive index and the thickness of the defect layer were varied.

Gevorgyan and Harutyunyan [17] considered some properties of defect modes in chiral photonic crystals with an anisotropic defect layer. They solved the problem by the Ambartsumian's layer addition method previously developed for the solution of astrophysical problems of multiple scattering in turbid media, and further extended to optical wave propagation through inhomogeneous media [18]. As a result, the influence of the defect layer thickness variation and its location in the chiral photonic crystal on defect mode properties were investigated. However, the amplitudes of the reflected and transmitted fields related to the incident wave were expressed by using 2×2 Jones matrix method, in which both the effect of refraction and multiple reflections between plate interfaces (dielectric discontinuities) are neglected [19].

Kopp and Genack [7] considered photonic properties of a defect created by twisting one part of an anisotropic structure relative to the rest of the sample. The authors restricted their discussion to samples with a chiral twist of 90° in the center of the sample, which creates a photonic defect at a frequency in the center of the stopband. However, varying the chiral twist angle in the range 0–180° tunes the defect frequency from the low to the high-frequency band edge. The study demonstrated that twisting one part of a chiral photonic structure about its helical axis creates a single circularly polarized localized mode that gives rise to an anomalous crossover in propagation. This is in contrast to a defect in a binary layered medium that produces two orthogonally polarized degenerate localized modes.

In another study, Schmidtke et al. [13] used two layers of a highly cross-linked CLC polymer film to create a phase jump of 90° in the cholesteric structure. They observed experimentally enhanced fluorescence inside the photonic stopband and laser emission due to a photonic defect mode in a dye-doped cholesteric polymer network. The observed emission peak can be attributed to the localized defect mode generated by the phase jump of the cholesteric helix and it agrees with the simulation results of Kopp and Genack [7], who have theoretically predicted the presence of a defect mode in the center of the stopband.

In the study [20], the twist defect mode inducted by a discontinuity of the director rotation around the helix axis at an interface of two photopolymerized cholesteric liquid crystal layers has been experimentally demonstrated in the transmission spectrum. The existing self-organized helical structure of such composite film acts as a one-dimensional photonic crystal. Finally, the authors observed optically pumped laser action based on the twist defect mode.

Ha et al. [21] considered quasi-periodic Fibonaccian phase defects in single-pitched cholesteric liquid crystalline systems. Fundamental optical properties were discussed as functions of phase jump (orientational defect angles), unit lengths, and the orders of Fibonacci systems. The performed simulations of reflection spectra demonstrated simultaneous red, green, and blue reflections or multiple photonic bandgaps, and the color of reflections could be controlled by an orientational defect angle.

Motivated by the literature review presented above, in this study, we considered a defect in the form of an anisotropic layer with different thicknesses, different locations of the defect layer inside the analyzed CLC cell, as well as different total thicknesses of the investigated system. To model optical phenomena, we implemented the 4×4 matrix method which gives accurate numerical results and was used in many previous papers [9,20–31]. On the contrary to numerous former studies examining the defect modes in nondeformed CLCs, we also investigated the influence of an external electric field applied to the considered CLC optical system with a defect layer inside. This is important from a technological point of view since the investigated optical system can also be additionally tuned during operation by altering the director field of the CLC.

The study is organized as follows. Section 2 presents the considered optical system, i.e., CLC cell with a defect layer inside, sandwiched between two homogenous and isotropic conducting glass samples. Section 3 shortly summarizes the most essential steps in the computational method of the director field of the CLC deformed by an externally applied electric field and the 4×4 matrix calculation method. Some interesting numerical simulations of the reflectance obtained for different configurations of the system, different values of the parameters, as well as different director fields of the CLC are reported and discussed in Section 4. Lastly, a brief summary and the conclusions of the study are outlined in Section 5.

2. Model of the Considered Optical System

Generally, in a liquid crystal cell under a given boundary condition and an externally applied field, the liquid crystal is in the director field configuration that minimizes the total free energy of this system [32]. In other words, there is a critical amplitude of the electric field below which the internal elastic strength of the liquid crystal exceeds the electric forces; therefore, the system remains undeformed from its base state. When a relatively large external electric field is applied to the cell, the liquid crystal will reorient because of the dielectric interaction between the crystal and the applied field. This bifurcation field-induced reorientation of the liquid crystal is referred to as the Fréedericksz transition [32].

Generally, determining the director field in a cholesteric liquid crystal device is complicated due to the large number of material parameters that characterize it. The tendency of the director to take on a particular spatial distribution can be understood in terms of the well-known elastic continuum theory. For instance, in papers [33–35], the problem of the continuum theory for nematic and cholesteric liquid crystals is exhaustively reported, and, therefore, it is not considered in detail in this study. According to this theory, elastic forces arising from the molecular structure of the material resist any distortion to the director field from its equilibrium state. Such distortions can arise for a number of reasons, including interaction with external electric/magnetic fields, coupling with the material flow of the liquid, or from interactions with confining surfaces [36]. When subjected to competing influences, the director will adjust throughout the sample and, as a result, must be treated as both space- and time-dependent function.

In our analysis, we considered situations, where the electric field interacts with the liquid crystal deformation. In the case of positive dielectric anisotropy of the CLCs, the molecules tend to align parallel to the electric field if voltage (electric field) is applied across the cell. As boundary conditions, we assumed the strong anchoring of the molecules of the CLC at surfaces of substrates and the deformation of the CLC characterized by the so-called tilt angle and twist angle of the molecules, which depends on the applied electric field, the anisotropy of the electrical permittivity of the liquid crystal, and the anisotropy of the elastic stiffness of the liquid crystal. The anisotropic defect layer is manifested by a locally constant director in the area of the defect layer.

Motivated by the structure of the simplest electro-optic devices, also in our system, a thin layer of a cholesteric liquid crystal is sandwiched between two plates (substrates) which are treated in a way which leads to the surface interaction imposing a specified orientation on the liquid crystal (boundary conditions of the director field). The optical properties of the investigated system can be additionally adjusted by applying an external electric field whose distorting influence on the CLC opposes that of the surfaces.

Figure 1 presents two configurations of the structure of the modeled optical system, i.e., CLC cell with a defect layer (DL) inside, and sandwiched between two (lower and upper substrates) homogenous isotropic media. Figure 1a presents the nondeformed director field, whereas Figure 1b shows the deformed director field caused by the applied electric field. The studied system is located in the right-handed Cartesian coordinate system, where the z-axis was chosen to be the twist axis of the CLC.

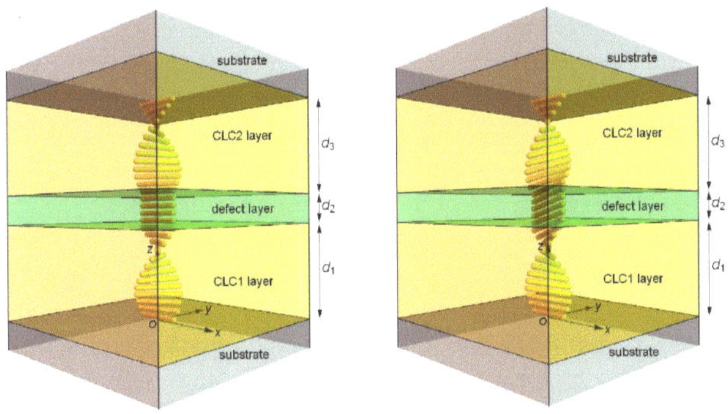

(a) (b)

Figure 1. A structure of the modeled cholesteric liquid crystal with a defect layer inside that is sandwiched between isotropic substrates, with the orientation of the director: (**a**) before the electric field is applied; (**b**) after the electric field is applied. The anisotropic defect layer is manifested by a locally constant director field in the range of the coordinate z from $z = d_1$ to $z = d_1 + d_2$.

The considered system can be treated as a multilayer system CLC1-DL-CLC2 with thicknesses of d_1, d_2, and d_3, respectively. In all considered cases, the whole CLC cell with the total thickness $D = d_1 + d_2 + d_3$ was divided into N equal multiple elementary layers (sliced into a large number of slabs) of the thickness 1 nm, which was parallel to the substrates (i.e., xy plane). Each imaginary elementary n-th layer can be treated as a homogenous layer with the same orientation of the optical axis, characterized by two angles, i.e., tilt angle and twist angle. The tilt angle was measured from the z-axis to the direction of the optical axis of the molecule, while the twist angle was measured from the x-axis to the projection of the optical axis of the molecule on the xy plane. As a result, each elementary layer can be treated locally as an anisotropic homogenous medium. The spatial period of the helical structure of the CLC characterized by the pitch p can be easily varied by adding dopants and changing temperature. The ordinary and extraordinary refractive indices were $n_o = 1.4639$ and $n_e = 1.5133$, respectively, whereas the pitch of the right-handed periodic helical structure was $p = 420$ nm. These values corresponded to the CLC cholesteryl-nonanoate–cholesteryl-chloride–cholesterylacetate (20:15:6) composition at the temperature 298 K [37]. As a result, the light with right circular polarization and normally incident onto a single CLC layer had a bandwidth in the range of $\lambda \sim (615–635)$ nm, whereas the light with the left circular polarization did not have any. The considered system was surrounded by the optical media with the refraction index n_s on its both sides, equal to $n_s = 1.4$ (the value that corresponds to glass medium). We assumed that both CLC and DL were not absorbing, and the values of the parameters n_s, n_o, and n_e did not depend on the wavelength of the incident light.

The incident light propagated in the lower substrate ($n = 0$), then in the CLC1, DL, and CLC2 ($n = 1, 2, \ldots, N$), and finally in the upper substrate ($n = N + 1$). The optical axis of the molecules of cholesteric liquid crystals laid on the xy plane which was parallel to the substrates and changed periodically along the z-direction. The orientation of the wave vector of the oblique incident nonpolarized beam of monochromatic light with wavelength λ in the lower substrate was represented by angles θ and ϕ. The angle θ was measured from the z-axis to the direction of the wave vector of the incident beam of light, while the angle ϕ was measured from the x-axis to the projection of the director of the wave vector of the incident beam of light on the xy plane [38,39].

3. Calculation Methods

We considered a cholesteric liquid crystal confined between parallel plates at $z = 0$ and $z = D$ and subject to strong planar anchoring at these surfaces as boundary conditions. We applied electric potential between the plates giving rise to an electric field parallel to the z-axis. We took into account the liquid crystal with positive electric anisotropy; therefore, the director tended to align parallel to the local electric field, and at the same time, the director was in competition with the direction of the strong planar anchoring. The deformation of the director occurred in two planes, in which it was possible to describe the director by two angles, namely the abovementioned tilt angle θ_c (measured from the z-axis to the direction of the optical axis of the molecule), and the twist angle ϕ_c (measured from the x-axis to the projection of the optical axis of the molecule on the xy plane).

For our calculations, the equations governing the static behavior of the liquid crystal with the variables θ_c and ϕ_c of the director field could be obtained from the equations expressing conservation of momentum, angular momentum, and Maxwell equations for the electric displacement field, with the appropriate boundary conditions that reflected the assumption of strong planar anchoring and that the voltage dropped across the considered cell was equal to the applied voltage. The solution of these equations in the nondimensional form depended on nondimensional parameters, i.e., the nondimensional elastic anisotropy, the nondimensional electric anisotropy, and the nondimensional applied voltage. In turn, the abovementioned nondimensional parameters depended on a large number of material parameters that characterized the investigated system. However, from the point of view of the optical properties of the system, such as reflectivity or transmissivity, the spatial distribution of the director had fundamental importance. Therefore, in our research, during the analysis of the considered optical cell in an external electric field, we took into account directly different distributions of the director field that corresponded to different amplitudes E of the applied electric field. In this way, the obtained results were more general and they could be directly adopted to systems with different material properties, both mechanical and optical.

To model optical phenomena in the analyzed system, we used the 4×4 matrix method implemented based on the mathematical formalism presented in [40]. More detailed information regarding the realized computer algorithm of the applied computation technique was presented in detail in one of our previous papers [39]. Below, only the main stages of this method are listed:

- Entering values of all required parameters of the considered optical system, i.e., n_s, n_o, n_e, d_1, d_2, d_3, p, λ, θ, and ϕ;
- Calculation of the dielectric tensors $\varepsilon(n)$ of the substrates ($n = 0$ and $n = N + 1$) and the dielectric tensors $\varepsilon(n)$ of the individual elementary crystal/defect layers ($n = 1, 2, \ldots, N$);
- Calculation of the wave vectors $\mathbf{k}_\sigma(n)$ of all four elementary waves ($\sigma = 1, 2, 3, 4$) in all elementary layers based on the wave equation;

- Computation of the optical polarization vectors $\mathbf{p}_\sigma(n)$ of the individual elementary waves representing the directions of the electric field as well as the corresponding vectors $\mathbf{q}_\sigma(n)$ of the magnetic field;
- Computation of the 4×4 transition matrix, which describes the relationships between amplitudes A_s, A_p of the electric field of the incident waves, amplitudes B_s, B_p of the electric field of the reflected waves, and amplitudes C_s, C_p of the electric field of the transmitted waves, both for s and p waves, respectively;
- Calculation of the reflection coefficient R (the fraction of the electromagnetic energy reflected) or optionally the transmission coefficient T (the fraction of the electromagnetic energy transmitted) of the incident light.

This mathematical approach is an effective technique in describing the optical properties of multi-layered structures and has been widely employed by many researchers in numerous problems, including the investigation of liquid crystal displays [41], the propagation of electromagnetic waves in stratified bianisotropic chiral structures [42], the study of various uniaxial, biaxial dielectric, magnetic materials [43,44], the investigation of the effect of stress-induced anisotropy on localized mode in photonic crystal [31], and the study of different kinds of cells of CLC lasers [45].

4. Results and Discussion

First, we examined the optical system for $d_1 = d_3 = 10\ p$ (further referred to as *model 1*) and for different values of the thickness d_2. In this case, the defect layer was located in the center of the studied cell. A beam of a monochromatic nonpolarized light was incident normally on the surface of the investigated optical system, i.e., perpendicular to the xy plane. For the sake of completeness and to increase the transparency of the obtained results, the results were presented in a few charts together with the curve obtained for the ideal system as a reference, i.e., without the defect layer (plotted by the dashed curve). The presented simulation results show that the thickness of the defect layer had a significant impact on the shape of the simulated spectral curves, especially in the band of the selective reflection. As one can see, visible dips in the spectral curves were induced by the introduction of the defect layer. Moreover, if the defect layer thickness was changed, the defect mode bandwidth of the incident light was changed too. Generally, the defect effect was present near the center and these curves had a certain asymmetry. However, for some values of the defect layer thickness, this effect was exactly in the center of the bandwidth and the spectral curves were symmetric. It should be noted that a similar effect can also be obtained in another way. For instance, a peak in transmission was observed in the middle of the bandgap, when a system with a spacerless defect was created in sculptured thin films by rotation of the substrate by an additional 90° in the middle of the film fabrication process [12].

The regularities in spectra presented in Figure 2 are easy to understand. When the thickness of the defect increased, the characteristic peak of the maximum value shifted toward the long-wave boundary of the bandwidth, while the second maximum appeared simultaneously on the left side of the spectral curve. For the layer thickness equal to $0.25\ p$ (Figure 2a), the spectral curve had a form of two symmetrical maxima with a slightly smaller amplitude around the wavelength of the maximum reflection for the system without the defect. With a further increase in the thickness of the defect (see Figure 2b), a similar situation occurred, but on the other side of the band gap (i.e., mirror image in comparison to the results presented in Figure 2a). For $d_2 = 0.5\ p$, the obtained spectral curve was almost identical to the curve obtained for the system without the defect (only a very small narrowing of the selective reflection bandwidth was visible). Increasing the defect thickness from $0.5\ p$ to p (see Figure 2c,d) caused a similar effect to changing the value of the d_2 parameter in the range from 0 to $0.5\ p$ (compare these curves with the curves presented in Figure 2a,b, respectively).

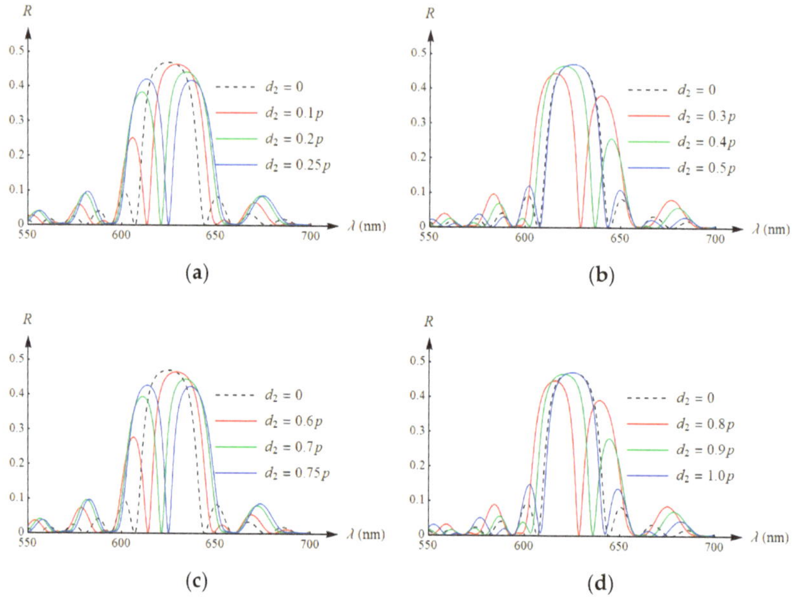

Figure 2. Reflection spectra for different values of the thickness d_2 of the defect layer (*model 1*): (**a**) d_2 equal to 0.1 p, 0.2 p and 0.25 p; (**b**) d_2 equal to 0.3 p, 0.4 p and 0.5 p; (**c**) d_2 equal to 0.6 p, 0.7 p and 0.75 p; (**d**) d_2 equal to 0.8 p, 0.9 p and 1.0 p.

Figure 3a shows a comparison of curves obtained for the defect layer thickness of 0.5 p, p, 1.5 p, and 2 p, while Figure 3b shows the comparison for d_2 = 0.25 p, 0.75 p, 1.25 p, and 1.75 p, respectively. In the first case (Figure 3a), we obtained similar results to for the system without the defect layer. The increase in the thickness of the defect layer (i.e., d_2 equal to 0.5 p, p, 1.5 p, or 2 p) caused only a very small and gradual narrowing of the bandwidth and a significant increase in the amplitude of the side peaks. In turn, for the value of the parameter d_2 tested in Figure 3b (i.e., d_2 equal to 0.25 p, 0.75 p, 1.25 p, and 1.75 p), two symmetrical peaks were obtained with a slightly smaller amplitude and with a zero reflection coefficient for $\lambda = (n_o + n_e) \cdot p/2 \approx 625$ nm. This gradual increase in the defect thickness resulted in a gradual but slight flattening of these spectral curves.

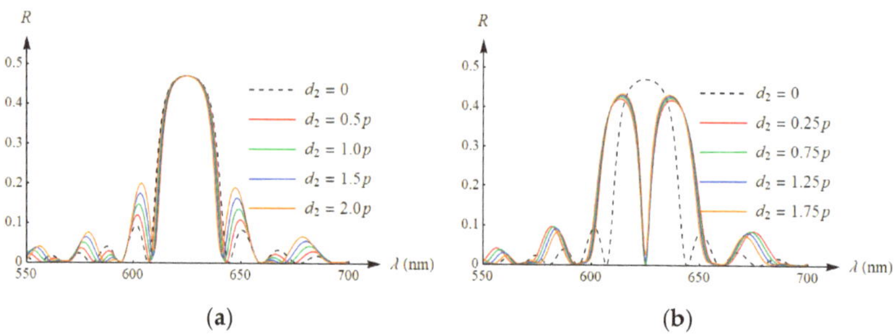

Figure 3. Reflection spectra for different values of the thickness d_2 of the defect layer (*model 1*): (**a**) d_2 equal to 0.5 p, 1.0 p, 1.5 p and 2.0 p; (**b**) d_2 equal to 0.25 p, 0.75 p, 1.25 p and 1.75 p.

Figures 4 and 5 show curves similar to those seen in Figures 2 and 3, but for the optical system with $d_1 = d_3 = 20\,p$ and different values of the thickness d_2 (further referred to as *model 2*). Through the analysis of the results of numerical simulations for the considered *model 2*, one can come to the same conclusions as for *model 1*. However, for greater values of the total thickness of the considered system, the observed selective reflection phenomenon was more pronounced. In the case of *model 1*, we observed a significant expansion of the selective reflection bandwidth (two adjacent peaks), with the simultaneous reduction in the average amplitude of the reflection coefficient. On the other hand, for *model 2*, the bandwidth extension was slightly smaller, but there was no significant reduction in the reflection coefficient in the bandwidth. This effect resulted from a larger value of the total twist angle of the liquid crystal optical axis, due to which the interference effects inside the crystal was stronger, and the selective reflection coefficient was greater.

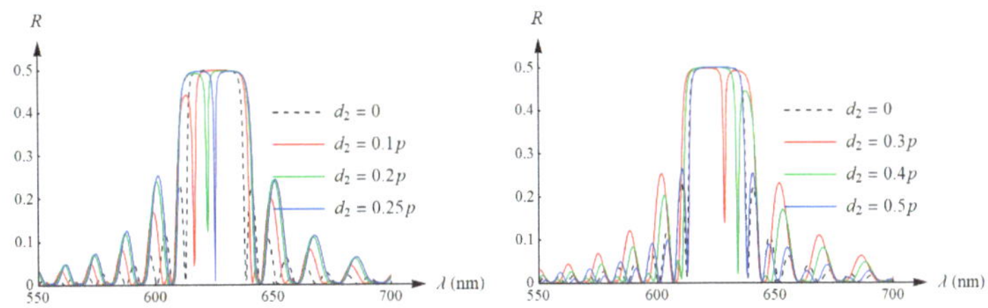

(a) (b)

Figure 4. Reflection spectra for different values of the thickness d_2 of the defect layer (*model 2*): (**a**) d_2 equal to 0.1 p, 0.2 p and 0.25 p; (**b**) d_2 equal to 0.3 p, 0.4 p and 0.5 p.

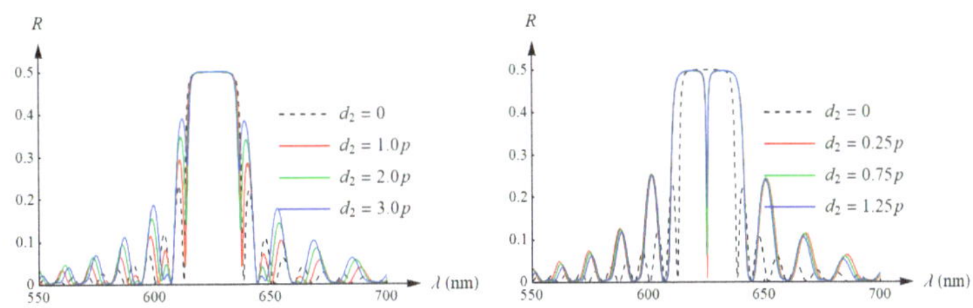

(a) (b)

Figure 5. Reflection spectra for different values of the thickness d_2 of the defect layer (*model 2*): (**a**) d_2 equal to 1.0 p, 2.0 p and 3.0 p; (**b**) d_2 equal to 0.25 p, 0.75 p and 1.25 p.

It is known that in the case of CLCs without defects, the value of the reflection coefficient is almost the same within the entire forbidden band λ in the range ($n_o p$, $n_e p$). Figure 6 shows the relationship between the reflection coefficient for the wavelength $\lambda = (n_o + n_e)\cdot p/2 \approx 625$ nm and the thickness of the defect layer, for both *model 1* and *model 2*. In addition, we also presented similar curves obtained for other wavelengths, i.e., $\lambda = n_o\cdot p \approx 615$ nm and $\lambda = n_e\cdot p \approx 635$ nm.

The simulation results presented in Figure 6 show that the reflection coefficient for individual wavelengths varied significantly depending on the thickness of the defect layer (in the range from 0 to 0.5), which was already observed in previous figures. Figure 6a shows that for the wavelength of 625 nm, the total reflection occurred for $d_2 = 0$, 0.5 p and p, while no reflection occurred for $d_2 = 0.25$ p and 0.75 p. Similar changes occurred for other wavelengths, i.e., 615 and 635 nm, with their maximum and minimum values occurring for other thicknesses of the defect layer. More interesting results are presented in Figure 6b. They were obtained for a greater total thickness of the investigated optical system. In addition, in this case, for the wavelength 625 nm and $d_2 = 0.25$ p, there was practically complete quenching of the selective reflection. However, for $d_2 = 0.75$ p, the value of the reflection coefficient no longer reached zero. This effect, also for larger values of the d_2 parameter (i.e., $d_2 = 1.25$ p), is clearly visible in Figure 5b. As the value of the thickness d_2 increased, the values of the reflection coefficient for $\lambda = 615$ and $\lambda = 635$ nm also increased. As a result, it can be stated that for the thicker tested optical system and thicker defect layers, the selective reflection bandwidth was significantly broadened without an undesired decrease in its amplitude.

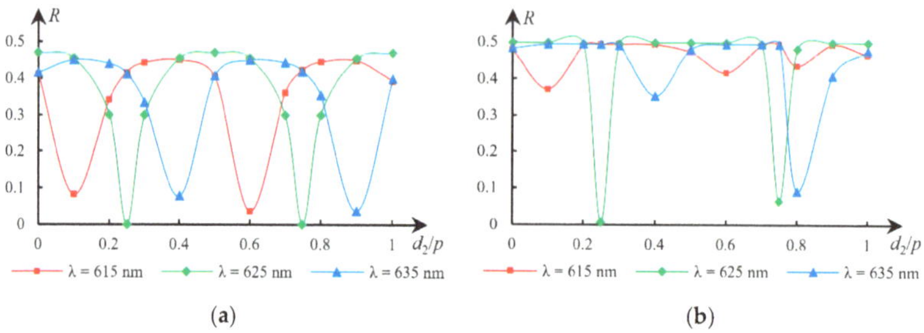

Figure 6. Reflection as a function of the thickness d_2 of the defect layer obtained for different values of the wavelength λ of the incident light: (**a**) *model 1*; (**b**) *model 2*. The curves interpolated based on the chosen computed values of the reflection represented by markers.

As it has been already noted, the abovementioned numerical tests were obtained for normal incidence of light. However, in our research, we also considered the oblique direction of the incident light. The direction of the beam of obliquely incident light can be described by two angles (θ and ϕ), and it was presented in detail in our previous papers [38,39]. Figure 7 presents color plots of the reflection coefficient R for the light with the wavelength $\lambda = 625$ nm incident onto *model 1* and for different values of the thickness d_2, whereas Figure 8 shows similar plots computed for *model 2*. Numerical simulations presented in Figures 7 and 8 show that the reflection coefficient practically did not change when the angle ϕ changed (especially for thicker layers of the investigated system (see Figure 8), while the distribution of the reflection coefficient changes when the θ angle changes. However, for the θ angle close to zero (almost perpendicular incidence of light), the reflection coefficient was practically constant. Significant changes could be observed only for large values of the θ angle, i.e., around 80–90°.

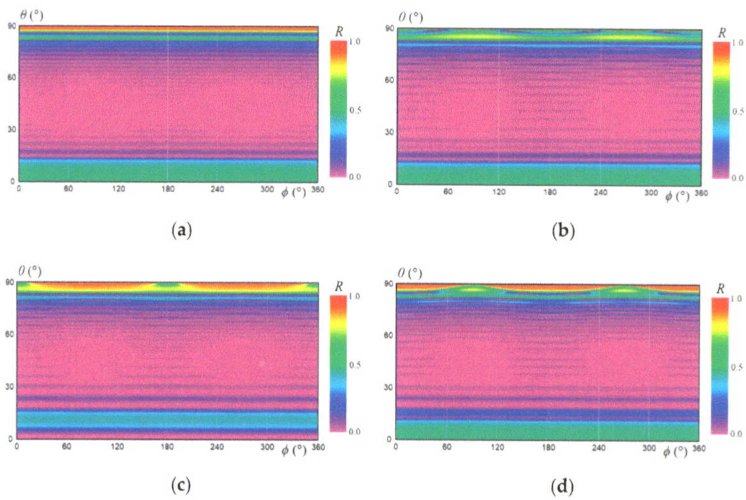

Figure 7. Color plots of the reflection coefficient R (*model 1*) obtained for different values of the thickness d_2 of the defect layer: (**a**) $d_2 = 0$; (**b**) $d_2 = 0.5\,p$; (**c**) $d_2 = 0.75\,p$; (**d**) $d_2 = 3\,p$.

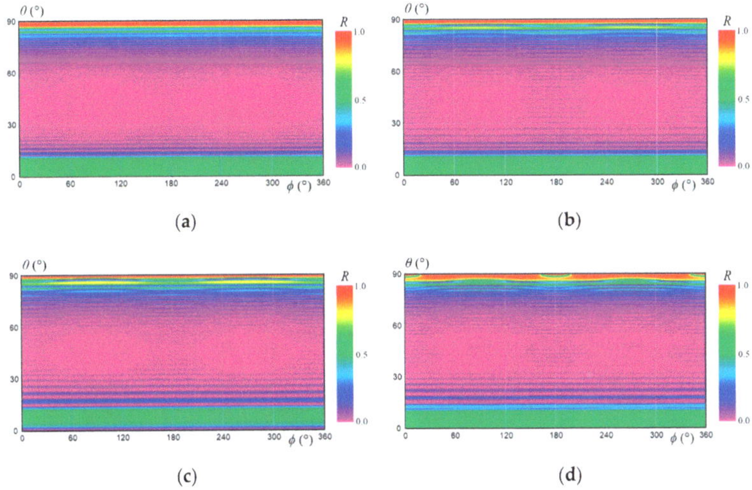

Figure 8. Color plots of the reflection coefficient R (*model 2*) obtained for different values of the thickness d_2 of the defect layer: (**a**) $d_2 = 0$; (**b**) $d_2 = 0.5\,p$; (**c**) $d_2 = 0.75\,p$; (**d**) $d_2 = 3\,p$.

The above figures present the simulations obtained for symmetrical systems ($d_1 = d_3$). Here we also examined the influence of the location of the defect layer inside the investigated CLC cell. To do it, we compared previously tested *model 2* (i.e., $d_1 = d_3 = 20\,p$, further referred to as *case A*), with two examples of the asymmetrical location of the defect layer, namely $d_1 = 10\,p$, $d_3 = 30\,p$ (further referred to as *case B*) and $d_1 = 30\,p$, $d_3 = 10\,p$ (further referred as *case C*). Again, we considered the case of normal incidence of light first. For comparison, the presented figures also show the case where there was no defect layer (marked as "without DL"). The analysis of the results presented in Figure 9 leads to the conclusion that in asymmetrical systems (i.e., *case B* and *case C*), the spectral curves differed slightly from the curves obtained for the symmetrical system (*case A*). In addition, the influence of the defect location in the reflection spectra for *case B* and *case C* was practically unobservable. Namely, we did not find a significant difference in spectral characteristics between *case B* and *case C*.

We also calculated the reflectance distribution (color plots) for the oblique incident light for all three cases (i.e., *case A*, *case B*, and *case C*). Color plots presenting these calculations in a graphical form are depicted in Figure 10. The presented results (here calculated for $d_2 = 0.1\,p$) confirmed the previous conclusions obtained on the basis of spectral curves and distributions of the reflection coefficient. Practically, no differences in the reflection spectra were observed for *case B* and *case C*, which is clearly shown in Figure 10d.

At the end of our research, we analyzed our optical system with a deformed director distribution caused by an externally applied electric field. As it was mentioned in Section 3, from the point of view of the optical properties of the examined system, the spatial distribution of the director field is fundamentally important. Therefore, in the present study, we directly took into account the different director distributions that corresponded to different amplitudes E of the electric field, without considering material parameters of the examined system. This approach was sufficient for the planned qualitative analysis.

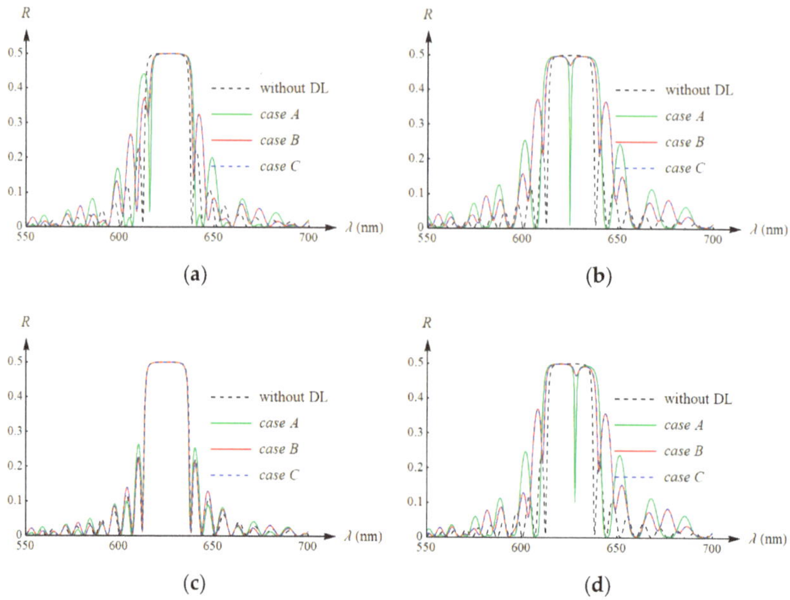

Figure 9. Reflection spectra (*model 2*) for different locations of the defect layer and different values of the thickness d_2: (**a**) $d_2 = 0.1\ p$; (**b**) $d_2 = 0.25\ p$; (**c**) $d_2 = 0.5\ p$; (**d**) $d_2 = 0.8\ p$.

Figure 11 presents the applied distributions of the director characterized by the tilt and twist angles as functions of the coordinate z. For better clarity of the presented results, Figure 11b shows the relative deformations of the angle, i.e., ϕ_c–ϕ_{c0}, in relation to the angle ϕ_{c0} obtained for the case of the undeformed director field. The presented curves correspond to different values of the amplitude E of the applied electric field, where the following relation was preserved: $E_6 > E_5 > E_4 > E_3 > E_2 > E_1 > E_0 = 0$. To increase the transparency of the obtained curves representing the director field, the results were presented together with the curve obtained for the undeformed director field as a reference, i.e., for the electric field $E_0 = 0$ (plotted by the dashed black curve). As it can be seen, the curves obtained for E_1, E_2, and E_3 were characterized by a relatively low director field deformation, while the curves obtained for E_4, E_5, and E_6 were characterized by a relatively large director field deformation. Therefore, the electric fields with the amplitudes E_1, E_2, and E_3 can be regarded as low electric fields, whereas electric fields with the amplitudes E_4, E_5, and E_6 can be regarded as high electric fields.

For further analysis, we considered only the optical system marked as *model 2*. Figures 12–14 show the spectral reflectance curves obtained for normal incidence of light, when the director field was deformed by an electric field with different amplitudes. The results shown in Figure 12 were obtained for the system without a defect layer ($d_2 = 0$), and the results depicted in Figure 13 were obtained for $d_2 = 0.25\ p$, whereas Figure 14 presents the results calculated for $d_2 = 0.5\ p$. Moreover, Figure 12a, Figure 13a, and Figure 14a correspond to the director field deformations obtained for the low electric field (E_1, E_2, and E_3), while Figure 12b, Figure 13b, and Figure 14b correspond to the director field deformations obtained for the high electric field (E_4, E_5, and E_6). The same results as those depicted in Figures 12–14 are shown again in Figure 15. However, Figure 15 directly shows the influence of different values of the thickness of the defect layer for a fixed director field, i.e., under the given electric field.

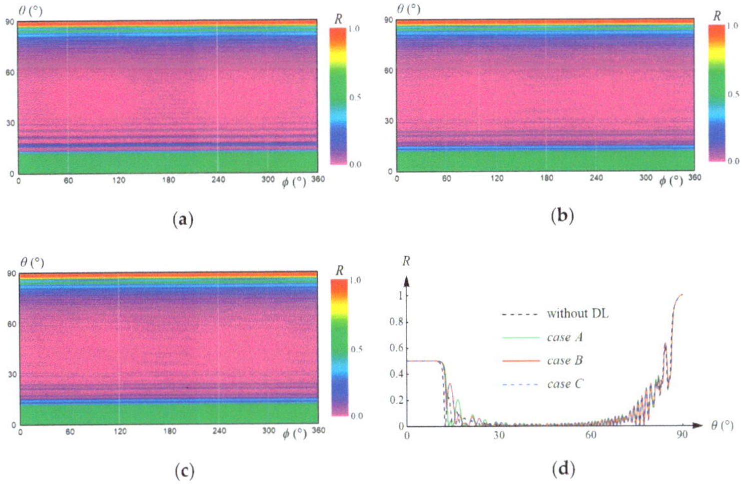

Figure 10. Color plots of the reflection coefficient (*model 2*) obtained for $d_2 = 0.1\,p$ and different locations of the defect layer: (**a**) *case A*; (**b**) *case B*; (**c**) *case C*. (**d**) Reflection as a function of the angle θ computed for $\phi = 0$ and different locations of the defect layer.

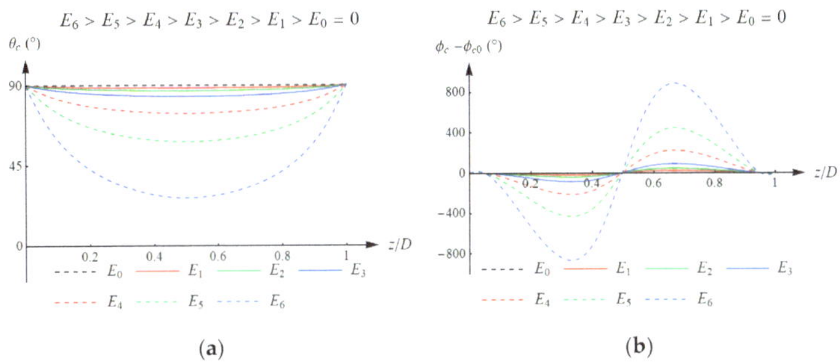

Figure 11. Director field deformations obtained for different values of the amplitudes of the applied electric field: (**a**) the tilt angle θ_c; (**b**) the relative deformation of the angle ϕ_c–ϕ_{c0} in relation to the angle ϕ_{c0} obtained for the case of the undeformed director field.

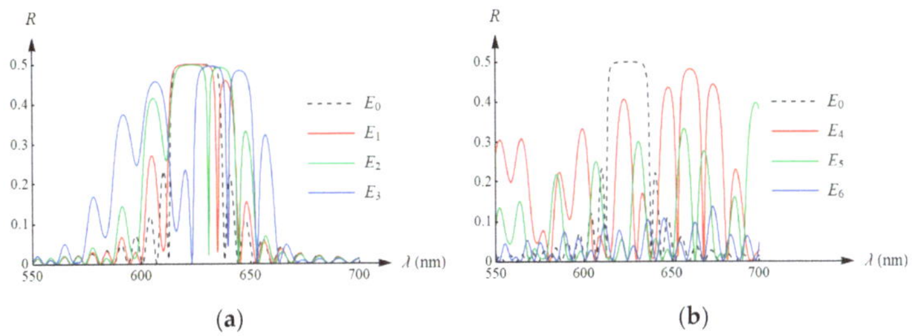

Figure 12. Reflection spectra of *model 2* without the defect layer ($d_2 = 0$) calculated for different values of the amplitude of the external applied electric field: (**a**) low electric field with the amplitudes E_1, E_2, and E_3; (**b**) high electric field with the amplitudes E_4, E_5, and E_6.

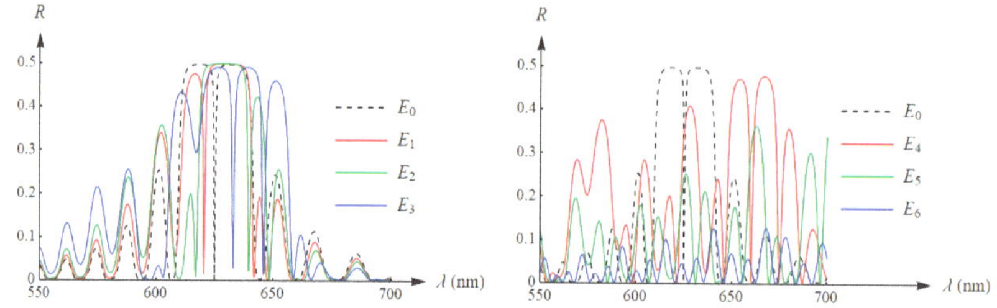

(a) (b)

Figure 13. Reflection spectra of *model 2* with a defect layer $d_2 = 0.25\,p$ calculated for different values of the amplitude of the external applied electric field: (**a**) low electric field with the amplitudes E_1, E_2, and E_3; (**b**) high electric field with the amplitudes E_4, E_5, and E_6.

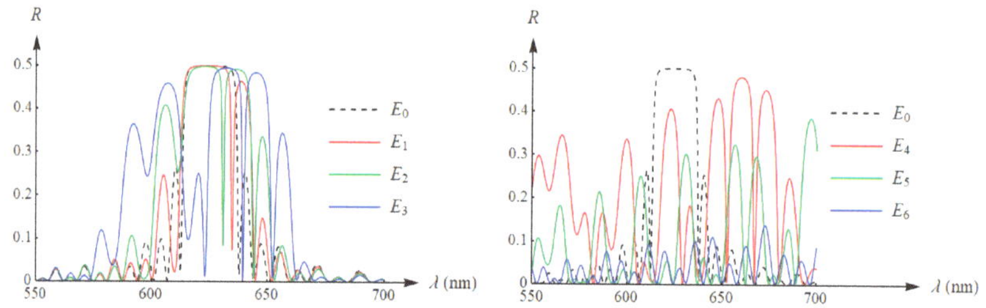

(a) (b)

Figure 14. Reflection spectra of *model 2* with a defect layer $d_2 = 0.5\,p$ calculated for different values of the amplitude of the external applied electric field: (**a**) low electric field with the amplitudes E_1, E_2, and E_3; (**b**) high electric field with the amplitudes E_4, E_5, and E_6.

In the case of the system without a defect layer, the application of a relatively low electric field caused a dip in the spectral curve near the right edge of the bandgap (see Figure 12a). With the increase in the amplitude of the electric field, the existing dip became more visible and shifted toward the short-wave boundary of the bandwidth. As a result, the applied electric field gave a similar effect to the introduction of a defect layer in the case of an undeformed director field of the CLC. What is more, the position of this dip can be precisely tuned within the range of the bandgap that exists for the undeformed director field. Similar conclusions can also be outlined when considering the results presented in Figure 14a that correspond to the system with a defect layer equal to 0.5 p. For the system with the defect layer thickness equal to 0.25 p, a dip in the spectral curve occurred in each presented case (see Figure 13a). For an undeformed director field, this dip was observed in the middle of the bandgap. The application of a low external electric field changed the position of this dip within the bandgap. It is worth noting that the shift of this dip can be both in the direction of the long-wave and the short-wave boundary of the bandwidth, depending on the amplitude of the applied electric field. The observation allowed the change in position of this narrow range. For example, this property can be useful when it is necessary to compensate the position of the dip due to the appearance of undesired external stimuli, such as mechanical or thermal stresses, acting on the considered system. In turn, the analysis of the results presented in Figure 12b, Figure 13b, Figure 14b clearly shows that in a high electric field applied to the tested system with or without a defect layer, the system lost the properties of the selective reflection and the defect modes. Finally, the analysis of the results presented in Figure 15 led to similar conclusions. In particular, for a fixed value of the electric field, the spectral curves were similar for both the system with no defect and the half-wavelength defect layer. In turn, the spectral curves obtained for the quarter-wavelength defect layer differed significantly from the spectral curves obtained for the system without the defect and with the half-wavelength defect layer, even if the applied electric field had the same value.

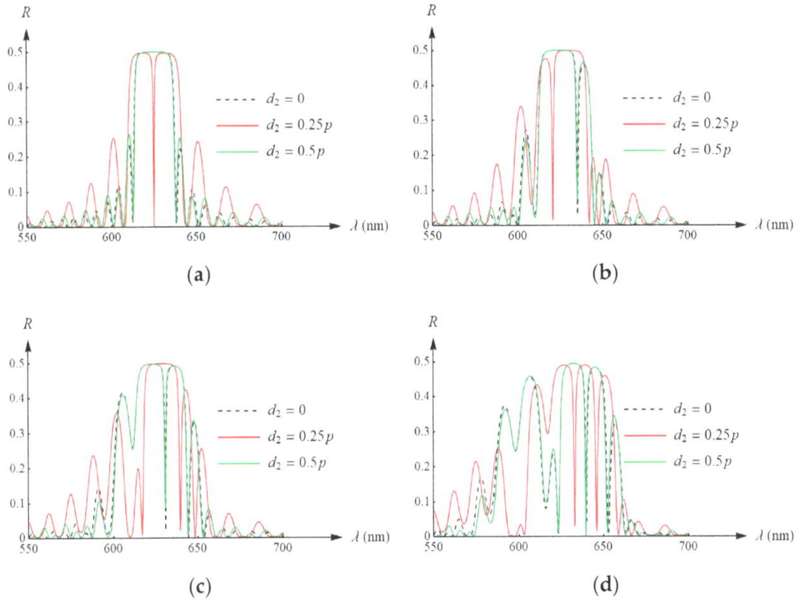

Figure 15. Reflection spectra of *model 2* for different thickness of the defect layer and different values of the amplitude of the external applied electric field: (**a**) $E = E_0$; (**b**) $E = E_1$; (**c**) $E = E_2$; (**d**) $E = E_3$.

The results of the present research may be of significant practical importance, since the considered model of the optical system is often found in engineering applications. Namely, in real engineering applications, CLCs are usually sandwiched between two parallel substrates coated with transparent conductive electrode, and voltage is applied across the cell. The applications of such systems can be greatly extended if the reflection can be electrically adjusted. This problem was considered in previous papers. As an example, Yang et al. [46] developed a (polymer network/nematic liquid crystal/chiral dopant) composite exhibiting the bandwidth of the selective reflection spectrum, which is wider and narrower reversibly with increasing and decreasing temperature, respectively. Lu and Chien [47] reported a method of fabricating single-layer color cholesteric liquid crystal displays from a polymer-stabilized cholesteric liquid crystal, where the reflective wavelength can be electrically switched to reflect blue or green from a cell initially reflecting the red color. Bailey et al. [48] investigated the electrical tuning of negative dielectric anisotropy cholesteric liquid crystals under the influence of AC and DC electric fields. They showed an electric effect with potential applications in tunable optical filters, optical pressure/stress sensors, or tunable lasers. In recent years, the application of polymer stabilization in CLCs provides a new dimension to achieve color tuning (for instance, see paper [8] and references cited therein). In the abovementioned paper [8], the authors experimentally investigated the broadening of the reflection band caused by an external electric field applied to the CLC cell. They studied a variety of factors affecting the reflection band broadening of polymer stabilized cholesteric liquid crystals, including the effect of moisture, the effect of alignment layer thickness, the effect of ionic additives, the relation between reflection band broadening and impedance, and the influence of morphology of the polymer network. As a result, by controlling these factors, they observed a wide reflection band to cover the entire visible region under low voltages.

The simulation results presented above concern the analysis of the CLC cell with a defect layer inside and a director field deformed by an external applied electric field. In these studies, we assumed that the spatial director's field distribution could be easily changed by an electric field. In general, studying the dynamic behavior of the considered cholesteric structures in an electric field is not an easy task, and some theoretical aspects in this area can be found in paper [6]. It should also be emphasized that the theoretical stability analysis of the orientation pattern of cholesteric liquid crystals both in electric and magnetic fields has been considered for a few decades. For instance, the unstable behavior of the planar texture between two conducting glass samples at voltages much below those needed to unwind the cholesteric helix was noticed [49]. More intensively, the problem of obtaining a stable and homogeneous uniform lying helix structure has been reported since the late eighties. However, most recently, the developed cholesteric liquid crystal device with the uniform lying helix structure shows stable and reversible characteristics with a short time response [50]. As a result, this enabled the solving of the long-standing problem of stability and hysteresis particularly to meet the manufacturing requirements dedicated to engineering applications.

The presented simulations regarding the appearance of similar defect modes agree with the related theoretical, numerical, or experimental results presented in the literature [51,52], where amplitudes and spectral positions can be elastically tuned. Moreover, our research leads to analogical conclusions as in other references, where it has been demonstrated that the amplitude, the number, and the position of the optical defect modes induced in the bandwidth depend on the thickness, the refractive index, and the anisotropy of the defect layer [17,53–57]. The proposed combination of the study of the defect modes caused by the defect layer inside and the tuning of the optical properties of the system by means of an electric field extends the existing research found in the literature and increases the possibility of its potential real applications.

5. Conclusions

First, we investigated the influence of an existing defect in the form of an anisotropic layer sandwiched by CLC. Numerical tests were carried out for different thicknesses of the system, different thicknesses of the defect layer, and different locations of this defect inside the investigated cell. In addition, we also studied the influence of an external electric field applied to the considered optical system that is important from a viewpoint of the technological aspects. Some interesting numerical simulations were obtained, illustrated, and discussed. The results were presented in the form of the reflection spectra for the normal incidence of light or color planar distributions of the reflection coefficient calculated for the oblique incident light. We showed that a defect layer placed inside CLC has a significant influence on the selective reflection phenomenon, and both the shape and the bandwidth of the spectral curve can be tuned by a dopant defect layer. In particular, the results obtained for *model 2* showed that defect modes can have an extremely narrow spectral width (narrow dip in the reflection spectrum or narrow peak in the transmission spectrum) on the order of a few Å.

The considered optical system serves as a model for more complex and realistic cholesteric liquid crystal devices. The obtained numerical results give a new insight into the behavior of practical devices based on the CLC with defect layers and additionally tuned by the external electric field. The analysis presented here demonstrates that there is a significant benefit for the industry to be gained from applying similar methods to models of liquid crystals in more general settings.

The research presented in this paper reflects current issues related to the new approach of modern optics, in which optical properties are determined by geometric factors, including dopants. An example of such a successful implementation is the investigated CLC cell with a defect inside. For instance, such a defect layer with a controlled anisotropy can be created by introducing a layer of a nematic liquid crystal, in which the orientation of the optical axis can be controlled electrically. In other cases, the thickness of the defect layer can be adjusted by the change of the applied voltage. This anisotropic defect provides an additional degree of freedom for tuning the defect modes. As a result, the analyzed CLC with a defect layer can work as an optical diode, which transmits light in one direction and prevents its propagation in the opposite direction. Currently, such research goals seem justified in engineering due to the expected conversion of electric signals to optical ones.

When it comes to using an external electric field in the considered system, the simplest operating regime is to switch between two states by the sudden application or removal of an electric field. In this way, in the absence of an external electric field, the analyzed system shows a selective reflection, including defect modes resulting from the existing defect layers. By means of small changes in the electric field, it is possible to precisely tune the optical properties of the system, which gives a similar effect to introducing defect layers with different thicknesses. In turn, the application of a high electric field causes the abovementioned optical properties to vanish.

In conclusion, in many technological applications, active tuning and switching of defect modes is desired. By introducing defect layers and applying an external electric field, the propagation of light can be precisely controlled through tailoring of the periodic dielectric structure of the CLC.

Author Contributions: Conceptualization, D.G. and J.A.; methodology and software, D.G.; investigation, validation and formal fnalysis, D.G. and J.A.; writing—original draft preparation and editing, D.G.; supervision and project administration, D.G. and J.A. All authors have read and agreed to the published version of the manuscript.

Funding: This research received no external funding.

Acknowledgments: We thank to Olga Jarzyna from Department of Automation, Biomechanics and Mechatronics of the Lodz University of Technology for technical support regarding preparation of this manuscript.

Conflicts of Interest: The authors declare no conflict of interest.

References

1. Venkatanarayanan, A.; Spain, E. 13.03—Review of recent developments in sensing materials. In *Comprehensive Materials Processing, Proceedings of the Reference Module in Materials Science and Materials Engineering*; Elsevier: Amsterdam, The Netherlands, 2014; Volume 13, pp. 47–101. [CrossRef]
2. Ramakrishnan, M.; Rajan, G.; Semenova, Y.; Farrell, G. Overview of fiber optic sensor technologies for strain/temperature sensing applications in composite materials. *Sensors* **2016**, *16*, 99. [CrossRef] [PubMed]
3. Beiu, R.-M.; Beiu, V.; Duma, V.-F. Fiber optic mechanical deformation sensors employing perpendicular photonic crystals. *Opt. Express* **2017**, *25*, 23388–23398. [CrossRef] [PubMed]
4. Li, G.; Zhang, S.; Zentgraf, T. Nonlinear photonic metasurfaces. *Nat. Rev. Mater.* **2017**, *2*, 17010. [CrossRef]
5. Beiu, R.-M.; Beiu, V.; Duma, V.-F. Fundamentals and Biomedical Applications of Photonic Crystals: An Overview. In Proceedings of the SPIE 10831, Seventh International Conference on Lasers in Medicine, Timisoara, Romania, 10 August 2018. [CrossRef]
6. Tarasov, O.S.; Krekhov, A.P.; Kramer, L. Dynamics of cholesteric structures in an electric field. *Phys. Rev. E* **2003**, *68*, 031708. [CrossRef]
7. Kopp, V.I.; Genack, A.Z. Twist defect in chiral photonic structures. *Phys. Rev. Lett.* **2002**, *89*. [CrossRef]
8. Nemati, H.; Liu, S.; Moheghi, A.; Tondiglia, V.P.; Lee, K.M.; Bunning, T.J.; Yang, D.-K. Enhanced reflection band broadening in polymer stabilized cholesteric liquid crystals with negative dielectric anisotropy. *J. Mol. Liq.* **2018**, *267*, 120–126. [CrossRef]
9. Kopp, V.I.; Bose, R.; Genack, A.Z. Transmission through chiral twist defects in anisotropic periodic structures. *Opt. Lett.* **2003**, *28*, 349–351. [CrossRef]
10. Yang, Y.-C.; Kee, C.-S.; Kim, J.-E.; Park, H.Y. Photonic defect modes of cholesteric liquid crystals. *Phys. Rev. E* **1999**, *60*, 6852–6854. [CrossRef]
11. Yablonovitch, E.; Gmitter, T.J.; Meade, R.D.; Rappe, A.M.; Brommer, K.D.; Joannopoulos, J.D. Donor and acceptor modes in photonic band structure. *Phys. Rev. Lett.* **1991**, *67*, 3380–3383. [CrossRef]
12. Hodgkinson, I.J.; Wu, Q.H.; Thorn, K.E.; Lakhtakia, A.; McCall, M.W. Spacerless circular-polarization spectral-hole filters using chiral sculptured thin films: Theory and experiment. *Opt. Commun.* **2000**, *184*, 57–66. [CrossRef]
13. Schmidtke, J.; Stille, W.; Finkelmann, H. Defect mode emission of a dye doped cholesteric polymer network. *Phys. Rev. Lett.* **2003**, *90*, 083902. [CrossRef]
14. Yokoyama, H.; Suzuki, M.; Nambu, Y. Spontaneous emission and laser oscillation properties of microcavities containing a dye solution. *Appl. Phys. Lett.* **1991**, *58*, 2598–2600. [CrossRef]

15. Painter, O.; Lee, R.K.; Scherer, A.; Yariv, A.; O'Brien, J.D.; Dapkus, P.D.; Kim, I. Two-dimensional photonic band-gap defect mode laser. *Science* **1999**, *284*, 1819–1821. [CrossRef]
16. Belyakov, V. *Diffraction Optics of Complex—Structured Periodic Media. Localized Optical Modes of Spiral Media*, 2nd ed.; Springer Nature Switzerland AG ©: Cham, Switzerland, 2019.
17. Gevorgyan, A.H.; Harutyunyan, M.Z. Chiral photonic crystals with an anisotropic defect layer. *Phys. Rev. E* **2007**, *76*, 031701-1–031701-9. [CrossRef]
18. Gevorgyan, A.A.; Papoyan, K.V.; Pikichyan, O.V. Reflection and transmission of light by cholesteric liquid crystal-glass-cholesteric liquid crystal and cholesteric liquid crystal(1)-cholesteric crystal(2) systems. *Opt. Spectrosc.* **2000**, *88*, 586–593. [CrossRef]
19. Gu, C.; Yeh, P. Extended Jones matrix method and its application in the analysis of compensators for liquid crystal displays. *Displays* **1999**, *20*, 237–257. [CrossRef]
20. Ozaki, M.; Ozaki, R.; Matsui, T.; Yoshino, K. Twist-defect-mode lasing in photopolymerized cholesteric liquid crystal. *Jpn. J. Appl. Phys.* **2003**, *42*, 472–475. [CrossRef]
21. Ha, N.Y.; Takanishi, Y.; Ishikawa, K.; Takezoe, H. Simultaneous RGB reflections from single-pitched cholesteric liquid crystal films with Fibonaccian defects. *Opt. Express* **2007**, *15*, 1024–1029. [CrossRef]
22. Matsui, T.; Ozaki, M.; Yoshino, K. Tunable photonic defect modes in a cholesteric liquid crystal induced by optical deformation of helix. *Phys. Rev. E* **2004**, *69*, 061715. [CrossRef]
23. Chen, J.-Y.; Chen, L.W. Defect modes in a stacked structure of chiral photonic crystals. *Phys. Rev. E* **2005**, *71*, 061708. [CrossRef]
24. Song, M.H.; Ha, N.; Amemiya, K.; Park, B.; Takanishi, Y.; Ishikawa, K.; Wu, J.; Nishimura, S.; Takezo, H. Defect-mode lasing with lowered threshold in a three-layered hetero-cholesteric liquid-crystal structure. *Adv. Mater.* **2006**, *18*, 193–197. [CrossRef]
25. Ozaki, R.; Matsuhisa, Y.; Yoshida, H.; Yoshino, K.; Ozaki, M. Optical properties and electric field enhancement in cholesteric liquid crystal containing different periodicities. *J. Appl. Phys.* **2006**, *100*, 023102. [CrossRef]
26. Ha, N.Y.; Jeong, S.M.; Nishimura, S.; Suzaki, G.; Ishikawa, K.; Takezoe, H. Simultaneous red, green, and blue lasing emissions in a single-pitched cholesteric liquid-crystal system. *Adv. Mater.* **2008**, *20*, 2503–2507. [CrossRef]
27. Nascimento, E.M.; Zanetti, F.M.; Lyra, M.L.; de Oliveira, I.N. Tunable reflectance spectra of multilayered cholesteric photonic structures with anisotropic defect layers. *Phys. Rev. E* **2010**, *81*, 031713. [CrossRef]
28. He, Z.; Ye, Z.; Cui, Q.; Zhu, J.; Gao, H.; Ling, Y.; Cui, H.; Lu, J.; Guo, X.; Su, Y. Reflection chromaticity of cholesteric liquid crystals with sandwiched periodical isotropic defect layers. *Opt. Commun.* **2011**, *284*, 4022–4027. [CrossRef]
29. Gevorgyan, A.H. Tunable reflectance of a two-defect-layer cholesteric liquid crystal. *Phys. Rev. E* **2011**, *83*, 011702. [CrossRef]
30. Vetrov, S.Y.; Pyatnov, M.V.; Timofeev, I.V. Photonic defect modes in a cholesteric liquid crystal with a resonant nanocomposite layer and a twist defect. *Phys. Rev. E* **2014**, *90*, 032505. [CrossRef]
31. Yang, O.; Wang, Y.; Wang, H.; Yanling, H.; Peng, J. Effect of stress-induced anisotropy on localized mode of one-dimensional photonic crystal with mirror symmetry. *Optik* **2015**, *126*, 5583–5586. [CrossRef]
32. Yang, D.-K.; Wu, S.-T. *Fundamentals of Liquid Crystal Devices*, 2nd ed.; John Wiley & Sons, Ltd.: Hoboken, NJ, USA, 2014. [CrossRef]
33. Ericksen, J.L. Equilibrium theory of liquid crystals. *Adv. Liq. Cryst.* **1976**, *2*, 233–298. [CrossRef]
34. Leslie, F.M. *Theory of Flow Phenomena in Nematic Liquid Crystals*; Ericksen, J.L., Kinderlehrer, D., Eds.; Springer: New York, NY, USA, 1987. [CrossRef]
35. Leslie, F.M. Continuum theory for nematic liquid crystals. *Contin. Mech. Thermodyn.* **1992**, *4*, 167–175. [CrossRef]
36. Self, R.H.; Please, C.P.; Sluckin, T.J. Deformation of nematic liquid crystals in an electric field. *Eur. J. Appl. Math.* **2002**, *13*, 1–23. [CrossRef]
37. Gevorgyan, A.H. About defect mode peculiarities in cholesteric liquid crystals with anisotropic defect layer inside. *Optik* **2018**, *154*, 656–661. [CrossRef]
38. Grzelczyk, D.; Awrejcewicz, J. Reflectance and transmittance of cholesteric liquid crystal sandwiched between polarizers. In *Springer Proceedings in Mathematics & Statistics*; Awrejcewicz, J., Ed.; Springer: Cham, Switzerland, 2018; Volume 249, pp. 147–158. [CrossRef]

39. Grzelczyk, D.; Awrejcewicz, J. Calculation of reflectance and transmittance of optical birefringent networks based on cholesteric liquid crystals. *Latin Am. J. Solids Struct.* **2019**, *16*, 20. [CrossRef]
40. Yeh, P.; Gu, C. *Optics of Liquid Crystal Displays*; John Wiley and Sons: New York, NY, USA, 1999.
41. Chen, C.-J.; Lien, A.; Nathan, M.I. 4 × 4 matrix method for biaxial media and its application to liquid crystal displays. *Jpn. J. Appl. Phys.* **1996**, *35*, 1204–1207. [CrossRef]
42. Ivanov, O.V.; Sementsov, D.I. Light propagation in stratified chiral media. The 4 × 4 matrix method. *Crystallogr. Rep.* **2000**, *45*, 487–492. [CrossRef]
43. Abdulhalim, I. Analytic propagation matrix method for anisotropic magneto-optic layered media. *J. Opt. A Pure Appl. Opt.* **2000**, *2*, 557–564. [CrossRef]
44. Lakhtakia, A.; Reyes, A. Theory of electrically controlled exhibition of circular Bragg phenomenon by an obliquely excited structurally chiral material—Part 1: Axial dc electric field. *Optik* **2008**, *119*, 253–268. [CrossRef]
45. Ortega, J.; Folcia, C.L.; Etxebarria, J. Upgrading the performance of cholesteric liquid crystal lasers: Improvement margins and limitations. *Materials* **2018**, *11*, 5. [CrossRef]
46. Yang, H.; Mishima, K.; Matsuyama, K.; Hayashi, K.-I.; Kikuchi, H.; Kajiyama, T. Thermally bandwidth-controllable reflective polarizers from (polymer network/liquid crystal/chiral dopant) composites. *Appl. Phys. Lett.* **2003**, *82*, 2407–2409. [CrossRef]
47. Lu, S.-Y.; Chien, L.-C. A polymer-stabilized single-layer color cholesteric liquid crystal display with anisotropic reflection. *Appl. Phys. Lett.* **2007**, *91*, 131119. [CrossRef]
48. Bailey, C.A.; Tondiglia, V.P.; Natarajan, L.V.; Duning, M.M.; Bricker, R.L.; Sutherland, R.L.; White, T.J.; Durstock, M.F.; Bunning, T.J. Electromechanical tuning of cholesteric liquid crystals. *J. Appl. Phys.* **2010**, *107*, 013105. [CrossRef]
49. Helfrich, W. Deformation of cholesteric liquid crystals with low threshold voltage. *Appl. Phys. Lett.* **1970**, *17*, 531–532. [CrossRef]
50. Yip, W.C.; Welch, C.; Mehl, G.H.; Wilkinson, T.D. A cholesteric liquid crystal device having stable uniform lying helix structure. *J. Mol. Liq.* **2020**, *299*, 112141. [CrossRef]
51. Avendanno, C.G.; Martinez, D. Elastic dependence of defect modes in one-dimensional photonic crystals with a cholesteric elastomer slab. *Photonics Nanostruct. Fundam. Appl.* **2018**, *30*, 30–38. [CrossRef]
52. Matsuhisa, Y.; Ozaki, R. High Q defect mode and laser action in one-dimensional hybrid photonic crystal containing cholesteric liquid crystal. *Appl. Phys. Lett.* **2006**, *89*, 101109. [CrossRef]
53. Ozaki, R.; Matsui, T.; Ozaki, M.; Yoshino, K. Electro-tunable defect mode in one-dimensional periodic structure containing nematic liquid crystal as a defect layer. *Jpn. J. Appl. Phys.* **2002**, *41*, 1482–1484. [CrossRef]
54. Ozaki, R.; Ozaki, M.; Yoshino, K. Defect mode switching in one-dimensional photonic crystal with nematic liquid crystal as defect layer. *Jpn. J. Appl. Phys.* **2003**, *42*, 669–671. [CrossRef]
55. Ozaki, R.; Ozaki, M.; Yoshino, K. Defect mode in one-dimensional photonic crystal with in-plane switchable nematic liquid crystal defect layer. *Jpn. J. Appl. Phys.* **2004**, *43*, 1477–1479. [CrossRef]
56. Arkhipkin, V.G., Gunyakov, V.A.; Myslivets, S.A.; Gerasimov, V.P.; Zyryanov, V.Y.; Vetrov, S.Y.; Shabanov, V.F. One-dimensional photonic crystals with a planar oriented nematic layer: Temperature and angular dependence of the spectra of defect modes. *J. Exp. Theor. Phys.* **2008**, *106*, 388–398. [CrossRef]
57. Kumar, R.; Kumar, D.; Kushwaha, A.S.; Srivastava, S.K. Study of one-dimensional nanolayered graded photonic crystal consisting of birefringent and dielectric materials. *Photonics Nanostruct. Fundam. Appl.* **2018**, *28*, 20–31. [CrossRef]

photonics MDPI

Article

Frequency Dependence of a Piezo-Resistive Method for Pressure Measurements of Laser-Induced Shock Waves in Solids

Ricardo Gonzalez-Romero [1], Marija Strojnik [2], Guillermo Garcia-Torales [1,*] and Gilberto Gomez-Rosas [1]

[1] University Center for Exact Sciences and Engineering, University of Guadalajara, Av. Revolución No. 1500, C.P. 44430 Guadalajara, Jalisco, Mexico; jaimer.gonzalezr@academicos.udg.mx (R.G.-R.); depfis@cucei.udg.mx (G.G.-R.)

[2] Optical Research Center, A. P. 1-948, C.P. 37000 Leon, Guanajuato, Mexico; mstrojnik@gmail.com

* Correspondence: garcia.torales@academicos.udg.mx

Abstract: A shock wave is a mechanical high-pressure pulse that travels inside a medium with a full width at half-maximum of a few nanoseconds that may be induced with a high-power laser pulse. A piezo-resistive measurement method to determine the shock wave pressure has been widely employed even though there is inner inaccuracy in the calibration process. We are interested in developing a precise theoretical model of laser material processing for applications in material sciences that includes the frequency dependence of the electronic post processing. We show an approach to determine the correction factor to frequency response at a high frequency of a piezo-resistive experimental setup and the results of the pressure measurements obtained in this experimental setup. The theoretical and experimental work demonstrates the feasibility of piezo-resistive methods to measure a laser-induced shock wave pressure in the nanosecond range. The correction factor of the frequency dependence calibration allows the technique to be applied in different shock wave experiments.

Keywords: shock wave; strain gauge; laser pulse; signal conditioner; laser shock processing (LSP); laser shock wave (LSW)

1. Introduction

A laser shock wave (LSW) is a mechanical pressure wave in a gigapascal (GPa) range with a full width at half-maximum (FWHM) in the order of nanoseconds (ns), induced by a high-power-density laser pulse [1–4]. An LSW propagates through a medium at a velocity greater than the sound speed (Mach > 1). The medium might be a solid, a liquid, or a gas. Some physical variables, including pressure, temperature, and particle velocity, to list a few, change discontinuously as the LSW propagates [5]. While the LSW propagates in a solid medium, some material characteristics improve through the relaxation of the compressive residual stresses [6].

Measurements of the LSW pressure inside a solid under controlled conditions are essential to regulate the optimal processing conditions (no ringing, for example) that result in the highest continuous pressure shock event. These material treatment applications include applied material science for high-performance metals, as in the aerospace, automotive, health, and defense industries, among others [7,8].

Several successful techniques have been proposed to measure the pressure induced by the LSW in solids. They include some piezoelectric methods [4,9], techniques that incorporate optical fibers [10,11], and piezo-resistive methods [12,13]. Such techniques developed in the previous decades presented some technological limitations. Piezoelectric methods have a nonlinearity response that needs to be considered. Fiber optic methods have a limited sensitivity and pressure range, restricting low-pressure measurements. Among the techniques mentioned, the piezo-resistive method has demonstrated, with excellent results, its outstanding features including a wide detection range, linearity, and temporal response in the nanosecond range [14].

In piezo-resistive methods, the signal processing system has two stages: a Wheatstone bridge (WB) and an instrumentation amplifier (IA). A WB is used to provide a sensor linearization if the sensor requires it and impedance coupling to the second stage. Additionally, the WB can be calibrated as a null method and is coupled to the IA. In the second stage, an IA is used to set the gain and reduce the common-mode signals. This gain is frequency-dependent [15], and it decreases with increasing frequency. In previous work [16], we observe that this relation of the gain with respect to the frequency produces an error in the LSW pressure measurement accuracy.

In this work, we propose an approach to determine a correction factor of the measurement system frequency response at a high frequency of a piezo-resistive experimental setup. In addition, we present the results of the shock wave pressure measurements obtained in this experimental setup when applying the correction factor. According to this mathematical model, we may adjust the critical parameters required for the LSW optimized operation, including the minimization of energy use. Our experimental results demonstrate that we may use a piezo-resistive method to measure a laser-induced shock wave pressure in the nanosecond range.

The Principle of the Shock Wave Induction

Figure 1 presents the concept of induction of the LSW generation in solids. A lens, covered with an antireflection coating, focuses a high-power laser beam on a small surface sample area of 0.3 cm^2. Most of the laser beam's energy is absorbed in the sample, subject to its specific optical material properties, as absorption and reflectivity at the laser wavelength.

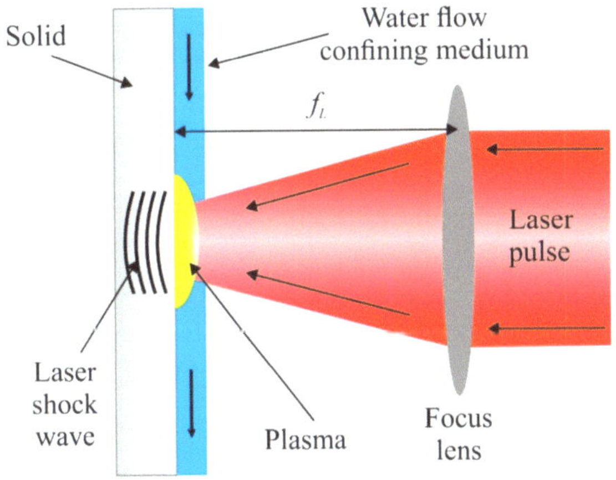

Figure 1. The physical principle of the laser shock wave induction. A high-power-density laser beam focused on the surface of a solid sample generates plasma. A high amplitude pressure wave with the magnitude of several GPa and a full width at half-maximum of a few nanoseconds is induced in the sample. It propagates from the sample's front surface to the backside while the plasma transfers its energy to the sample in an adiabatic process.

A thin layer of the sample surface is sublimated due to a rapid and large increase in the surface layer's temperature to about 10,000 K [17]. As described by Z. Zhang et al. [18], this process is separated into three stages: stage I, the absorption of laser pulse; stage II, the formation and expansion of laser-induced plasma; and stage III, the plastic deformation on the target surface due to the laser shock loading. We will focus on stages I and II, since stage III is beyond our study's scope.

Stage I begins after the laser pulse is focused on the solid surface. Electrons instantaneously absorb the photons' energy on the Fermi surface. This instant energy absorption thermalizes the electrons through high temperatures in the picosecond range. Thus, vibration on the lattice begins to transmit their energy through phonons' emissions. This energy exchange tends to thermal equilibrium after decades of picoseconds and is described by the two-temperature model of the thermal conservation of energy expressed in terms of the electron temperature diffusion:

$$C_e\frac{\partial T_e}{\partial t} = \nabla\cdot(\kappa_e\nabla T_e) - \Gamma_{ei}(T_e - T_i) - \left(\frac{\partial E_e}{\partial V} + p_e\right)\frac{\partial V}{\partial t} + R_{Abs} - R_{Emis} + S, \quad (1)$$

$$C_i\frac{\partial T_i}{\partial t} = \nabla\cdot(\kappa_i\nabla T_i) - \Gamma_{ei}(T_e - T_i) - \left(\frac{\partial E_i}{\partial V} + p_i\right)\frac{\partial V}{\partial t} - q\frac{\partial V}{\partial t}, \quad (2)$$

where the subscript e and i correspond to electrons and ions, respectively; $C_{e,i}$ are the heat capacities per unit volume; $T_{e,i}$ are the temperature; $E_{e,i}$ are specific internal energies; $\Gamma_{ei} = C_e Z^2 \ln\Lambda_{ei}\left(A^2 V T_e^{3/2}\right)$ is the electron-phonon coupling constant, which describes the energy exchange rate between electrons and ions where Z is the charge index, A is the relative atomic mass, and Λ_{ei} is the Coulomb term for electron–ion collisions; $p_{e,i}$ are the energy density; q is the Von-Neumann artificial viscosity; and V is the specific volume. Therefore, $\kappa_{e,i}$ are the thermal conductivities, which are described by

$$\kappa_{e,i} = 20\left(\frac{2}{\pi}\right)^{3/2}\frac{(k_B T_{e,i})^{5/2} k_B \tau_{e,i}}{m_e^{1/2} e^4 Z \ln\Lambda_{ei,ii}}, \quad (3)$$

where k_B is the Boltzmann constant, m_e is the electron mass, e is the electron charge, and Λ_{ii} is the Coulomb term for ion–ion collisions, and $\tau_i = 1$ and $\tau_e = 0.43Z/(3.44 + Z + 0.26 \ln Z)$. R_{Abs} and R_{Emis} are radiation absorption and emission terms, which are given by

$$R_{Abs} = c\sum_{g}^{N_F}\left(\sigma_g^{PA} E_{R,g}\right), \quad (4)$$

$$R_{Emis} = \frac{8\pi(k_B T_e)^4}{c^2 h^3}\sum_{g}^{N_F}\left(\sigma_g^{PE}\int_{x_g}^{x_{g+1}}\frac{x^3}{e^x + 1}dx\right), \quad (5)$$

where g is the frequency group index, N_F is the number of frequency groups, c is the speed of light, $E_{R,g}$ is the radiation energy density for group g, σ_g^{PA} and σ_g^{PE} are Planck mean opacities for emission and absorption, respectively, and $x = \hbar\omega/k_B T_e$ where h is Planck's constant, and ω is the angular frequency. Finally, S is the laser source in Equation (1), described by

$$S = \alpha(1 - R)I(t)\left(\frac{w_0}{w}\right)^2 e^{-2r^2/w^2} e^{-\alpha z}, \quad (6)$$

where $\alpha = 1/\delta$ is the absorption coefficient of the sample, which is determined by the skin-depth $\delta = 1/\sqrt{\pi\mu f\sigma_0}$ of the electromagnetic wave penetrating into the solid, where μ is the magnetic permeability, f is the frequency of the wave, and σ_0 is the electrical conductivity; R is the reflectivity of the sample surface; $I(t)$ is the laser intensity as a function of time; w_0 is the radius at the beam waist; w is the radius of the beam spot the size on the metal surface; and z is the depth measured perpendicular to the surface of the target. Equation (6) relates the intensity of the laser and the solid sample's optomechanical properties for the laser heating.

In stage II, even if the laser pulse's fluence is near tens of J/cm^2, it is not sufficiently high to ionize all material. The ionization observed at the end of the stage I is due to collisions between fast electrons, multiphoton processes, and an ionization field near the laser focus area. This ionization is not symmetrical due to the collisions' incident, and reflected angles are not equal under an electromagnetic field. This asymmetry leads to the rupture of the ionization field. However, in every collision between fast electrons and ions, there is energy gain in the form of the probability of the number of ionizations:

$$\left(\frac{dE_e}{dt}\right)_{gain} = \left(\frac{8k_B}{\pi}\right)^{1/2} \frac{ne^2 I}{m^{3/2}\varepsilon_0 c\omega^2}\sigma T_e^{1/2}, \tag{7}$$

where σ is the collision cross-section, n is the number density, and ε_0 is the vacuum permeability. If this energy gained from collisions is greater than the sum of the mean ionization energy or other energy dissipation components, an avalanche process induces a plasma breakdown. The plasma breakdown is described with a macroscopic two-fluid Vlasov model assuming the Maxwell-Boltzmann conditions to obtain the velocity moment, density, and pressure.

$$\frac{\partial f_\sigma}{\partial t} + \frac{\partial f_\sigma}{\partial \vec{x}} \cdot \frac{d\vec{x}}{dt} + \frac{\partial f_\sigma}{\partial \vec{v}} \cdot \frac{d\vec{v}}{dt} = \sum_\alpha C_{\sigma\alpha}(f_\sigma), \tag{8}$$

where the distribution function is denoted by $f(x,v,t)$ to characterize the instantaneous configuration of a large number of particles, such as the density of particles at each point in the phase spatial. If a one-dimensional (1D) approach is used, the conservation of mass equation with Lagrangian coordinates can be written as

$$\frac{\partial V}{\partial t} = V\frac{\partial u}{\partial r} = \frac{\partial u}{\partial m_0}, \tag{9}$$

where $V = 1/\rho$ is the specific volume, u is the velocity fluid, and m_0 is the Lagrangian mass variable. This mass density can be obtained by Equation (8) if the results are adapted to the number density of electrons and ions. If we assumed an isotropic system, the scalar pressure P_σ is obtained by multiplying Equation (8) with velocity and integrating it over the velocity space, taking into account the temporal evolution of the mean fluid velocity for each species (σ and α) under Lorentz force in the electromagnetic field $\vec{E}$ and $\vec{B}$:

$$n_\sigma m_\sigma \frac{\partial \vec{u}_\sigma}{\partial t} + n_\sigma m_\sigma \left(\vec{u}_\sigma \cdot \nabla\right)\vec{u}_\sigma = n_\sigma q_\sigma \left(\vec{E} + \vec{u}_\sigma \times \vec{B}\right) - \nabla P_\sigma - \vec{R}_{\sigma\alpha}, \tag{10}$$

where m_σ is the mass of the particle, q_σ is the particle charge, and $R_{\sigma\alpha}$ is the net frictional drag force due to collisions of species σ with species α. Furthermore, the momentum conservation equation is

$$\frac{\partial u}{\partial t} = -\frac{1}{\rho}\frac{\partial}{\partial r}(P+q) = -\frac{\partial}{\partial m_0}(P+q) + \dot{u}_{TN}, \tag{11}$$

$P = Pe + Pi + Pr$ is the total fluid pressure due to electrons, ions, and radiations; q is the von Neumann artificial viscosity; and u_{TN} is the velocity change momentum exchange from the slowing down of fast particles. If the Vlasov equation of Equation (8) is multiplied with a factor of $m_\sigma v^2/2$ and integrated over the N-dimensional velocity space, a so-called Vlasov second moment is defined as

$$\frac{N}{2}\frac{\partial P_\sigma}{\partial t} + \frac{N}{2}\vec{u}_\sigma \cdot \nabla P_\sigma + \frac{2+N}{2} P_\sigma \nabla \cdot \vec{u}_\sigma = -\nabla \cdot \vec{Q}_\sigma + \vec{R}_{\sigma\alpha} \cdot \vec{u}_\sigma - \left(\frac{\partial W}{\partial t}\right)_{E_{\sigma\alpha}}, \tag{12}$$

Equations (1), (2), (6), (10), and (12) describe the laser–matter interaction involved in the induction of LSW. The other steps in obtaining the equations that explain all procedures of laser–matter–plasma interaction are described in [18].

As a remark, the amount of the laser power needed to heat a thin surface layer of the sample to the point of sublimation depends on the full width at half-maximum (FWHM) of the laser pulse, the irradiated spot area, and the relative area of the irradiation with respect to the sample linear dimension [19]. The large amount of the energy absorbed in the thin surface layer transforms the state of matter into plasma. Consequently, the plasma transfers its energy to the solid in an adiabatic process. This process induces a mechanical, high-amplitude pressure wave of several GPa inside the sample with an FWHM of a few nanoseconds. The pressure wave travels through the material from the front surface called the impact zone, to the back surface. We determined that the FWHM of the LSW is about 50 ns in our experiment. This parameter was determined at a frequency of 20 MHz. We consider that the LSW initiates a fast change in pressure (~25 ns), and it achieves peak pressure at a time of about one-half of this interval (13 ns).

A dielectric layer, referred to as a confining medium, is used to confine the plasma, delay its expansion, and increase the LSW pressure. The confining medium may be chosen as air, water, or glass, to name a few [20]. In many cases, water is preferred due to practical industrial considerations. It is coupled perfectly to the geometry. It is replaced easily after a pulse, has a low cost, and produces acceptable confinement. Two arrangements of confining plasma with water are possible: immersion and flow. In the immersion confinement, the confining medium surrounds the sample, assuring a single medium between the sample and the incident laser pulse. However, the immersion type generates a collapsing bubble that induces a secondary shock wave in the sample. That problem is overcome using water that absorbs a fraction of the laser pulse energy. Thus, the flow type, or so-called water jet, does not have this disadvantage. The water layer is only a few millimeters thick. Therefore, the water flow is considered laminar, avoiding the creation of the cavitation bubble.

2. Materials and Methods

The experimental procedure to obtain LSW pressure within solids using the piezo-resistive detection method and correction factor consists of two parts: First, we measured the pressure of the LSW. Then, we found the frequency response of the second stage to determine the correction factor and the attenuation of the gain. We used two plate samples to induce an LSW and perform the requisite pressure measurements; one was a 5 mm aluminum 6061-T6, and the other was a 1.3 mm aluminum 6063-T5. The mechanical properties for both samples are shown in Table 1. In this study, we chose aluminum blocks without absorbent coating [21–23] since they are used in many applications where lightweight metals are required (for example, in the aerospace industry). However, any other material, titanium or iron, may be used in the future, applying our proposed method.

Table 1. Aluminum alloy mechanical properties.

Aluminum Alloy	Tensile Strength	Elastic Limit	Elongation	Young's Modulus	Fatigue Strength	Thermal Shock Resistance
	MPa	MPa	%	GPa	MPa	
6061-T6	290	241	10	69	96	14
6063-T5	160	97	11	68	70	8

2.1. Experimental Setup

Figure 2 shows the experimental setup's typical schematic layout to induce an LSW in a solid sample. The power source is a Q-switched (Quantel Brilliant B Nd:YAG laser) with a wavelength of 1064 nm and a spot diameter of about 10 mm. It is operated in a single pulse mode, for the duration of 6 ns, with an energy density equal to 0.77 J cm^{-2}. These settings of the laser power source are typical for laser shock processing (LSP). A high-reflectivity, dielectric-coated planar mirror *DM* turns the beam direction for compact layout and decreases back-reflection into the laser cavity [24]. A biconvex lens *L* covered with an antireflection coating, a focal length of f_L = 1 m, and a transmission of about 96% is used to focus the laser beam onto a spot diameter of about 1.2 mm on the sample surface. The local fluence on the laser spot is 68.08 J cm^{-2}. A fast silicon photodiode

(Thorlabs DET10A) is connected to trigger the oscilloscope. This photodiode has a spectral responsivity interval of 200–1100 nm and a 1 ns rise time. A water flow (in blue) confines the medium.

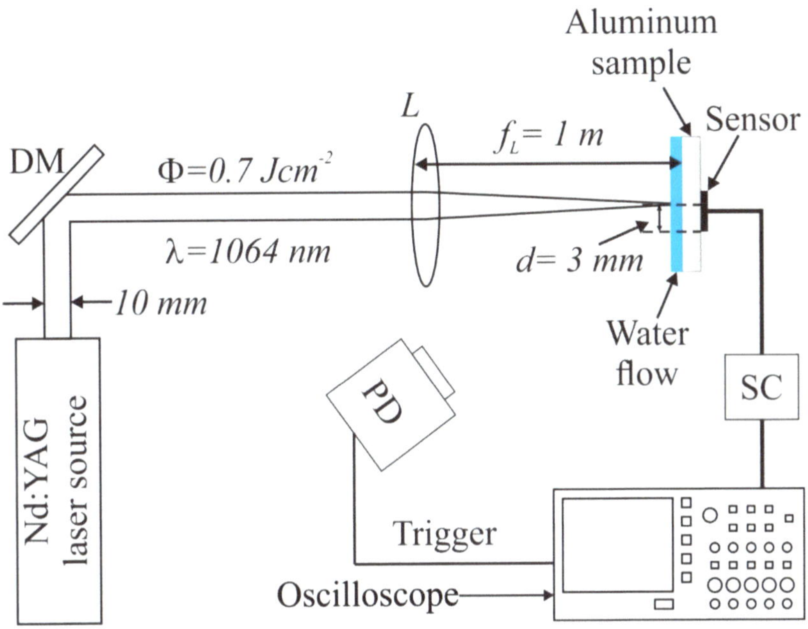

Figure 2. Experimental setup to induce a shock wave in a solid (aluminum, in our case). A single pulse from the Nd:YAG high power laser with λ = 1064 nm, an energy density of 0.77 J cm^{-2}, 6 ns pulse duration, and a spot diameter of 10 mm is focused on the sample surface with a bi-convex lens L covered with an antireflection coating to decrease a spot diameter to about 1.2 mm. A manganin piezo-resistive sensor is attached to the backside of the sample. The output signal from the sensor is connected to a signal conditioning system SC (WB, IA and a high-pass filter).

We measured the pressure induced by the shock wave through embedding a manganin piezo-resistive sensor (VISHAY LM-SS-125CH-048) in the back of the sample probe at a distance d = 3 mm below the laser spot, as shown in Figure 3. This distance corresponds to the length between the center and the boundary of the sensor active region. The sensor is attached by a thin film of bond M-Bond 200 (cyanoacrylate) for pressure gauges. This film is smaller (<2 µm) than the wavelength of the shock wave (320 µm for Al [25]), and its effects on the shockwave propagation may be negligible. Further, this alloy of the sensor has a quasi-linear response for wide pressure ranges, as shown by Duan Z. et al. [26]. Its output signal is connected to a quarter Wheatstone bridge transducer. Then, the signal is conditioned by an IA with a gain G = 670. The output signal is filtered with a first-order passive high-pass filter, with a cut-on frequency of 300 kHz to reduce the 1/f noise. The detector output signal is exhibited in an oscilloscope (Tektronix TDS 7054) that measures data with a 500 MHz bandwidth, time-gated at 2 *GS/s* and within a time window of 1 µs. Data is recorded for further processing on a PC.

We used a quarter Wheatstone bridge as a conditioning piezo-resistive sensor. The resistance of the sensing element (in our experiment, manganin) is denoted by $R_{Gauge} = R_{xNOM} + \Delta R_{Gauge}$, where $R_{xNOM} = 48\ \Omega$ is the nominal resistance with no disturbance and $\Delta R_{Gauge} = 0.1 \times 10^6 \times P \times S_S \times R_{xNOM}$ is the resistance change due to a shock wave transit (the steady-state of the sensor is $\Delta R_{Gauge} \neq 0$), where P is the shock wave pressure in [Bar] ($1\ Bar = 100 \times 10^3$ Pa) and S_S is sensor sensitivity in $[\Omega\Omega^{-1}(\text{kPa})^{-1}]$. Subsequently, an instrumentation amplifier (IA) is connected to the WB terminals to reduce the two inputs' common-mode signals. The output voltage V_{out} of the IA is related to the pressure P in terms of sensitivity S_s and gain $G(f)$ of the measurement system:

$$P(G(f), V_{out}) = 0.1 \times 10^6 [G(f) S_S R_{xNOM}]^{-1} \left[\left(\frac{R_3(R_2\alpha + \beta)}{R_1\alpha - \beta} \right) - R_{xNOM} \right], \tag{13}$$

where $\alpha = GV_S$ and $\beta = V_{out}(R_1 + R_2)$. The gain of the measurement system decreases with frequency. Thus, an attenuation factor $G(f)$ is included for the gain explicitly in the mathematical model. Its frequency response is experimentally measured in the following subsections. Appendix A shows the electronic details of the experiment and the mathematical manipulation to obtain Equation (13).

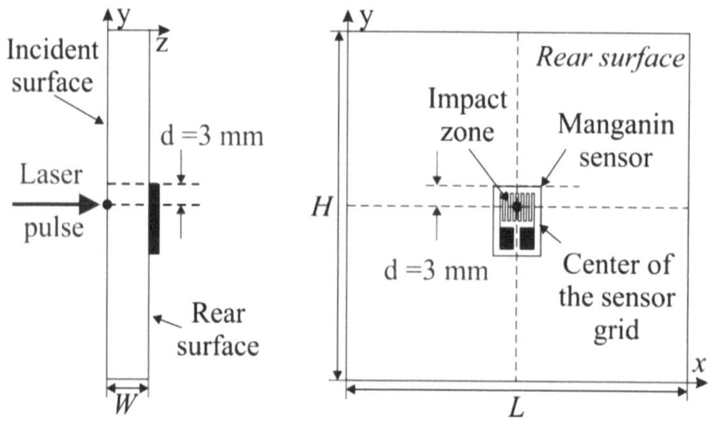

Figure 3. The manganin piezo-resistive sensor is embedded in the back of the solid sample. It is placed 3 mm below the impact zone, between the center and the active region boundary, at the middle of the sample.

2.2. Frequency Dependence Correction

The experimental setup to determine the correction factor $G(f)$ of the frequency dependence is exhibited in Figure 4. This correction factor is used to accurately determine the pressure signal in Equation (13). A sweep generator (Rode & Schwarz SWM05, 10–18,000 MHz) produces a sinusoidal signal that increases its frequency in a time window of 30 seconds, with an amplitude of 448 mV peak-to-peak. This sweep generator is used as the signal source to obtain a Bode diagram of the measurement system. The initial and the final window frequency of the signal source are 10 MHz and 30 MHz (20 MHz bandwidth), respectively. The output signal of the measurement system is recorded with a spectrum analyzer (HP8593A), with a 9–26.5 GHz bandwidth, a span bandwidth of 20 MHz, and the center frequency at 30 MHz. A Tektronix TDS 2024B oscilloscope is connected with a temporal base to observe the sweep signal. A Faraday cage with a cut-off frequency of 60 GHz is set all around the measurement system to minimize external noise and electrical interference. A cooling system has been implemented to reduce the thermal (Johnson) noise. The electrical characteristics of the measurement system are listed in Table 2.

Table 2. Electrical characteristics of the measurement system.

IA Voltage Source	WB Voltage Source	Output Impedance	Current Consumption
[Vdc]	[Vdc]	[Ω]	[mA]
±4	10	50	205

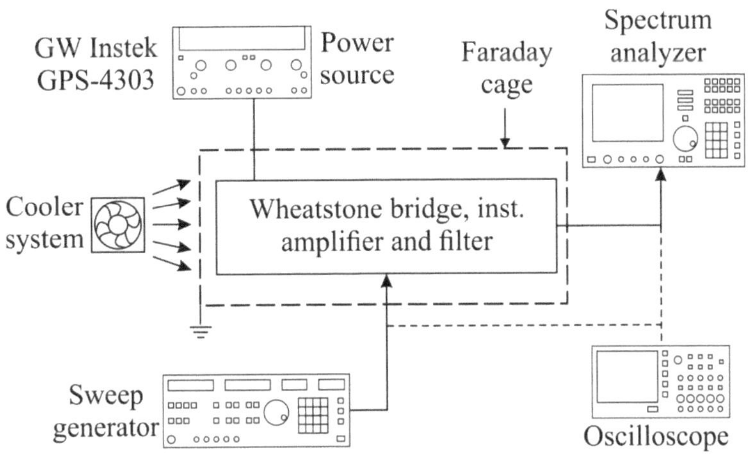

Figure 4. Functional pictographic diagram of the experimental setup to measure the system frequency response. A sweep generator produces a variable frequency sinusoidal signal probe to use as the input signal. A spectrum analyzer is connected to the output of the measurement system to find the frequency response curve, including its associated gain attenuation function.

3. Results and Discussions

3.1. Frequency Dependence Correction

The Bode diagram of the pressure measurement system is displayed in Figure 5a to determine the correction factor. The vertical axis corresponds to the gain of the measurement system with respect to the frequency. The usual response of the LSW in the experimental setup is expected in the frequency range of 10 MHz to 50 MHz. The gain decreases with frequency and may be approximately described as a cubic curve. Examining Figure 5a, we can read that it is about −27 dB at 20 MHz, corresponding to a correction factor *G(f)* = 0.0006. Dots represent a cubic polynomial fit, and a fourth-degree polynomial fit is shown with dashes, $G(f)_1$ and $G(f)_2$. Their corresponding functions, with frequency measured in MHz, are as follows:

$$G(f)_1 = -1.3f^3 + 7.5\times10^{-3}f^2 - 0.66f - 17, \tag{14}$$

$$G(f)_2 = -6.4\times10^{-6}f^4 + 0.76\times10^{-3}f^3 - 0.025f^2 - 0.12f - 20. \tag{15}$$

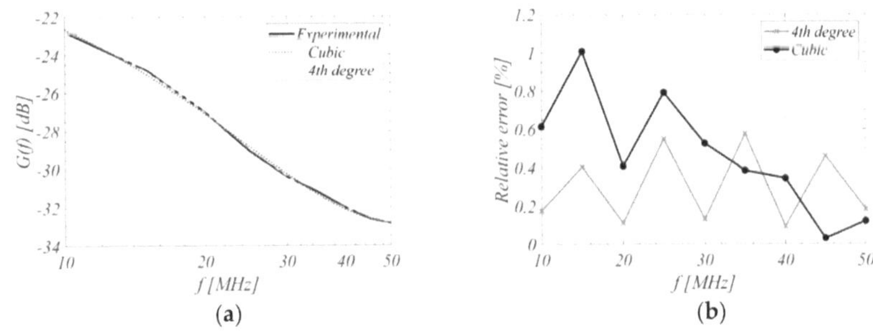

Figure 5. (**a**) Bode diagram of the frequency response of the pressure measurement system. Two polynomial fits were used to obtain a mathematical model for the frequency correction factor curve, a cubic and a quartic one. (**b**) The relative error of fourth- and third-degree polynomial fits. The fourth-order polynomial fit has a lower relative error than the cubic; however, the improvement is relatively small.

The relative error for each gain function is graphed in Figure 5b. We summarize, according to errors obtained in the polynomial fits, the fourth polynomial order fit is used in the mathematical model of the correction factor curve.

3.2. Shock Wave Pressure

We measured the parameters describing the behavior of the shock wave pressure in our experimental setup. The shock wave pressure was calculated using Equation (13) for two aluminum samples: 1.3-mm thick Al 6063-T5 and 5-mm thick Al 6061-T6. Figure 6 shows the shock wave pressure in both Al 6061-T6 5-mm thick slab (gray) and Al 6063-T5 1.3-mm thick slab (black) for a 1-μs time interval after the irradiation pulse.

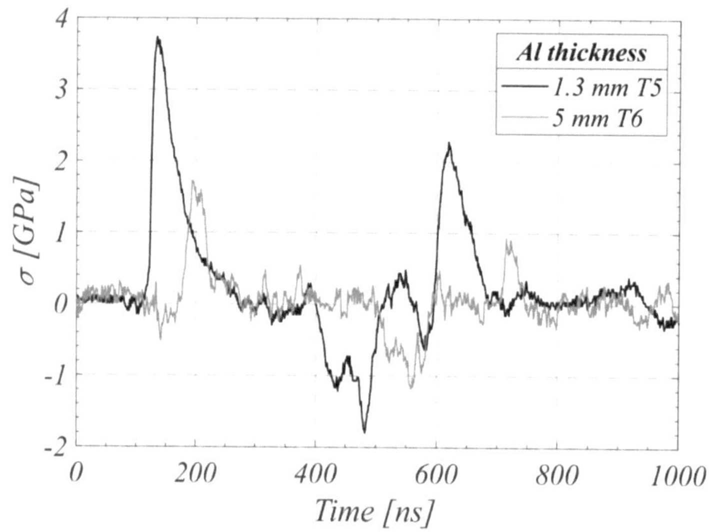

Figure 6. Shock wave pressure profile in both Al 6061-T6 5-mm thick (gray) and Al 6063-T5 1.3-mm thick (black) in a 1 μs time interval. It shows the two first pressure peaks. The second pressure peak represents one cycle, corresponding to the shock wave's travel from a sample rear surface to the sample front side. Peak pressure for the 1.3-mm thick sample is 3.8 GPa, and the 5-mm thick sample is 1.68 GPa. Similar profile tendencies are observed between the highest consecutive peaks.

The shock wave pressure in Al 6063-T5 is 3.8 GPa, and for Al 6061-T6 it is 1.68 GPa. The pressure peak recorded for the Al 6063-T5 sample is about two-times higher than that of the four-times thicker Al 6061-T6 sample. A higher amount of energy is dissipated after the shock wave is transmitted into a thicker slab of material, producing a decreased peak amplitude. This is interpreted as the whole block reacting to the shock imposed by the thermal impulse. The pressure difference between samples is due to the shock wave traveling a longer distance on the 5-mm thick specimen than the 1.3-mm thick specimen. In this more extended route, the shock wave undergoes a rapid pressure amplitude decrement due to dissipation in the form of heat. Moreover, in the 5 mm sample, the thermal shock resistance (shown in Table 1) causes a more noticeable pressure difference. The pressure difference does not have a linear relationship with the material thickness due to nonlinear shock wave propagation inside the medium. This is also in agreement with the results reported by H. Hu et al. [27].

Figure 6 presents a similar profile for the first and the second pressure peaks and an approximate mirror profile for the first compression peak. We refer to the positive peak as a pressure peak, and a negative peak is a compression peak. The difference between a pressure peak and a compression peak is their trajectories. We consider pressure as the wave traveling from the induction surface towards the sensor and compression as the wave traveling from the sensor towards the induction surface. The time elapsed between the first pressure peak at 140 ns and the second pressure peak at 620 ns for Al 6063-T5 sample of 480 ns is significantly lower than the time difference for Al 6061-T6 of 520 ns (200 ns, 720 ns).

We ascribe it to different slab thicknesses. The shock wave takes longer to reach the rear surface for the second time in Al 6061-T6 as compared to that in Al 6063-T5. The shock wave period is about 50 ns. In making this assessment, we consider the onset of the abrupt increase in pressure after a steady state.

We show the pressure signals displaced along the time scale to the onset of the first pressure pulse in Figure 7. The time delay between the beginning of the first pressure peak and the beginning of the first pressure valley for the 1.3-mm thick sample is 348 ns, and for the 5 mm sample, it is 365 ns. This is a time difference of 17 ns or 5% (negligible) between two samples. The time delay between the beginning of the first pressure valley and the beginning of the second top pressure peak is 136 ns for a 1.3-mm thick sample and 156 ns for a 5 mm sample. This is a time difference of 20 ns between two samples. Analyzing the beginning of the second pressure peak in samples, we can appreciate a different initial time. This difference in time is about 44 ns. These three time differences indicate a general increase in times between complete shock wave reflections cycles in the 5 mm sample to those in the 1.3 mm sample; that is, the shock wave propagation times increase between each reflection from the rear surface to the front surface. This difference in time arises because the aluminum slabs have different chemical compositions. The specimen Al 6061-T6 has an improved elasto-mechanical property and thermal shock resistance compared to those of aluminum 6063-T5, as shown in Table 1. This enhanced composition produces more shock wave dissipation, exhibited as a signal deceleration.

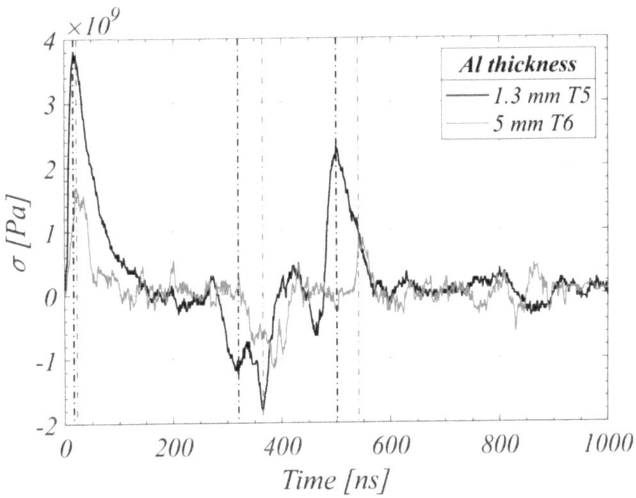

Figure 7. The peak pressure signals are displaced versus time from the beginning of the first pressure pulse. The difference in time between both samples, marked as vertical lines, at the beginning of the first pressure peak and origin of the first compression peak is 17 ns. At the beginning of the first compression peak and the beginning of the second pressure peak, the delay time is 20 ns. The difference between the beginning of the second pressure peak between both samples is 44 ns.

4. Conclusions

We demonstrate the signal-processing stages effects on the measurement quality, and its precision needs to be considered to measure the LSW pressure with a piezo-resistive method. We have shown that the manganin-based piezo-resistive method may be used to measure pressure in nanoseconds with fast response time, high sensitivity, linearity, and low cost.

We measured the pressure in two different aluminum alloy samples in thickness and composition, obtaining high-accuracy results. Energy dissipation and propagation of the shock wave exhibit a nonlinear relation to material thickness. Similarly, the time increase between complete reflection cycles in both samples has not indicated a linear relationship to the material thickness increases.

We presented and characterized a novel method to determine the frequency correction factor of the gain due to frequency dependence of a piezo-resistive measurement system for laser shock wave pressure measurements in nanoseconds, within 10 to 50 MHz, corresponding to a shock of 20 ns to 100 ns. We custom-designed the signal processing system to handle such rapid events, obtaining its correction factor curve.

We expanded the application of the piezo-resistive method to the laser-induced pressure measurement using a manganin sensor and a custom-designed signal processing system. This allows a systematic performance comparison for different frequencies, not only for the laser induced shock, but also for other shock wave phenomena.

Author Contributions: G.G.-T. and G.G.-R. conceived the ideas; R.G.-R. realized the prototype; R.G.-R. performed the measurements; all the authors analyzed the data; M.S. and G.G.-T. contributed to the principle verification, results, and discussion of the work; M.S. and R.G.-R. wrote the paper. All authors reviewed and commented on the manuscript; G.G.-T. provided supervision and guidance in this work. All authors have read and agreed to the published version of the manuscript.

Funding: Air Force Office of Scientific Research: FA9550-18-1-0454.

Data Availability Statement: The data presented in this study are available upon request from the corresponding author.

Acknowledgments: The author Gonzalez-Romero acknowledges the assistantship to the SNI Investigator level III funded by the CONACYT, as well as the Universidad de Guadalajara, for the support to accomplish this research.

Conflicts of Interest: The authors declare no conflict of interest.

Appendix A

We used a quarter Wheatstone bridge (WB) for our experiment, as exhibited in Figure A1. This electronic, passive-component device is used as a conditioning piezo-resistive sensor. The resistance of the sensing element (in our experiment, manganin) is denoted R_{Gauge}. The voltage V_{CD}, called the differential voltage, measured between terminals C and D, is different from zero as the pressure wave travels through the sensor, generated upon a change in sensor resistance.

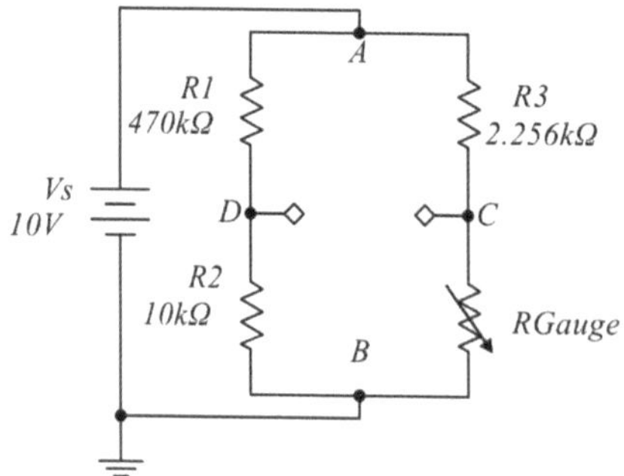

Figure A1. The Wheatstone bridge is an electronic, passive-component, null-detection sensor incorporated into the piezo-resistive sensors. The components R_1, R_2, R_3 are fixed-value resistors. The R_{Gauge} represents the manganin as the piezo-resistive gauge element. The incremental change in resistance of the piezo-resistive sensor due to the creation of the shock wave yields a differential voltage increase between points C and D.

The voltage V_{CD} of the bridge is obtained analytically using the Thevenin's theorem. The resistors R_1, R_2, and R_3 are selected in such a way to linearize the sensor responsivity. Linearity of the bridge is achieved by decreasing sensitivity with appropriated values of R_1/R_2 and R_3/R_{Gauge} [16]. The WB resistances are R_1 = 470 kΩ, R_2 = 10 kΩ, R_3 = 2.256 kΩ, and R_{Gauge} = 48 Ω (null detection). These values yield a ratio of 47 for R_1/R_2 and R_3/R_{Gauge} relationships. The increase in voltage V_{CD} may be related to ΔR_{Gauge} resistance.

$$V_{CD} = V_S\left(\frac{R_1}{R_1 + R_2} - \frac{R_3}{R_3 + R_{Gauge}}\right), \tag{A1}$$

where V_S is WB voltage source and $R_{Gauge} = R_{xNOM} + \Delta R_{Gauge}$ is the resistance change. R_{xNOM} = 48 Ω is the nominal resistance. If a shock wave disturbs the steady-state of the sensor ($\Delta R_{Gauge} \neq 0$), the resistance change ΔR_{Gauge} is defined as follows:

$$\Delta R_{Gauge} = 0.1 \times 10^6 \times P \times S_S \times R_{xNOM}, \tag{A2}$$

here, P is pressure in [Bar] (1 Bar = 100×10^3 Pa) and S_S is sensor sensitivity in $[\Omega\Omega^{-1}(\text{kPa})^{-1}]$. Subsequently, an instrumentation amplifier (IA), shown in Figure A2, is connected to the WB terminals C and D. Further signal processing and conditioning results in noise reduction and signal magnification. The instrumentation amplifier reduces common-mode signals of the two inputs. It has a gain obtained by $G = 1 + (2R_6/R_{Gain})$. The output voltage V_{out} corresponds to the differential voltage of the Wheatstone bridge V_{CD} multiplied by gain G

$$V_{out} = \Delta V_{CD} \times G. \tag{A3}$$

Solving for pressure P, we substitute Equation (A1), and Equation (A2) into Equation (A3), then we obtain pressure P in terms of sensitivity S_S of the measurement system, output voltage V_{out}, and system gain G:

$$P(G(f), V_{out}) = 0.1 \times 10^6 [G(f) S_S R_{xNOM}]^{-1}\left[\left(\frac{R_3(R_2\alpha + \beta)}{R_1\alpha - \beta}\right) - R_{xNOM}\right], \tag{A4}$$

here $\alpha = GV_S$ and $\beta = V_{out}(R_1 + R_2) + R_2$. The total gain of the measurement system decreases with frequency. Thus, an attenuation factor $G(f)$ is included for the gain explicitly in the mathematical model.

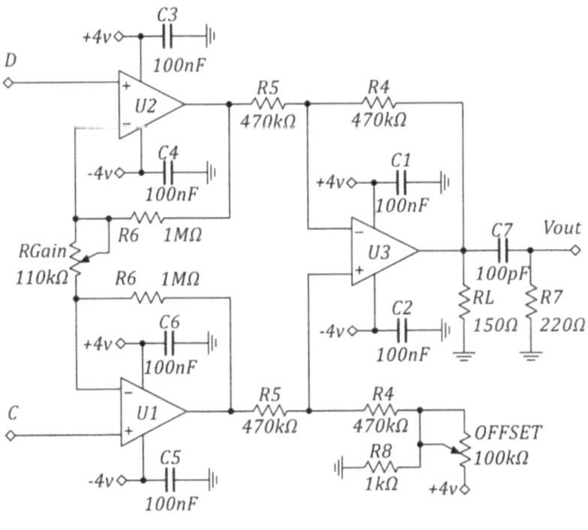

Figure A2. An instrumentation amplifier is used to subtract sensor signal, reject common mode voltage signals, and amplify the signal. A three-operational amplifier topology is implemented to obtain experimentally measured pressure data.

References

1. Romero, J.R.G.; Gomez-Rosas, G.; Garcia-Torales, G.; Strojnik, M. Piezoresistive Method for a Laser Induced Shock Wave Detection on Solids. In Proceedings of the SPIE Optical Engineering Applications, San Diego Convention Center, San Diego, CA, USA, 30 August 2017; Strojnik, M., Kirk, M.S., Eds.; SPIE: San Diego, CA, USA, 2017.
2. Berthe, L.; Fabbro, R.; Peyre, P.; Tollier, L.; Bartnicki, E. Shock waves from a water-confined laser-generated plasma. *J. Appl. Phys.* **1997**, *82*, 2826–2832. [CrossRef]
3. Fabbro, R.; Fournier, J.; Ballard, P.; Devaux, D.; Virmont, J. Physical study of laser-produced plasma in confined geometry. *J. Appl. Phys.* **1990**, *68*, 775–784. [CrossRef]
4. Peyre, P.; Berthe, L.; Fabbro, R.; Sollier, A. Experimental determination by PVDF and EMV techniques of shock amplitudes induced by 0.6–3 ns laser pulses in a confined regime with water. *J. Phys. D Appl. Phys.* **2000**, *33*, 498–503. [CrossRef]
5. Kiran, P.P.; Chelikani, L.; Bagchi, S.; Tewari, P.S. Direction-dependent asymmetric expansion of laser-induced shock waves in air. *Proc. SPIE* **2010**, *7721*, 772124.
6. Gomez-Rosas, G.; Rubio-Gonzalez, C.; Ocaña, J.L.; Molpeceres, C.; Porro, J.A.; Morales, M.; Casillas, F.J. Laser Shock Processing of 6061-T6 Al alloy with 1064 nm and 532 nm wavelengths. *Appl. Surf. Sci.* **2010**, *256*, 5828–5831. [CrossRef]
7. Montross, C.S.; Wei, T.; Ye, L.; Clark, G.; Mai, Y. Laser shock processing and its effects on microstructure and properties of metal alloys: A review. *Int. J. Fatigue* **2002**, *24*, 1021–1036. [CrossRef]
8. Gujba, A.K.; Mamoun, M. Laser peening process and its impact on materials properties in comparison with shot peening and ultrasonic impact peening. *Materials* **2014**, *7*, 7925–7974. [CrossRef]
9. Cao, Y.; Feng, A.; Hua, G. Influence of interaction parameters on laser shock wave induced dynamic strain on 7050 aluminum alloy surface. *J. Appl. Phys.* **2014**, *116*, 153105. [CrossRef]
10. Cranch, G.A.; Lunsford, R.; Grün, J.; Weaver, J.; Compton, S.; May, M.; Kostinski, N. Characterization of laser-driven shock waves in solids using a fiber optic pressure probe. *Appl. Opt.* **2013**, *52*, 7791–7796. [CrossRef] [PubMed]
11. Zou, X.; Wu, N.; Tian, Y.; Zhang, Y.; Fitek, J.; Maffeo, M.; Niezrecki, C.; Chen, J.; Wang, X. Ultrafast Fabry-Perot fiber-optic pressure sensors for multimedia blast event measurements. *Appl. Opt.* **2013**, *52*, 1248–1254. [CrossRef]
12. Rosenberg, Z.; Yaziv, D.; Partom, Y. Calibration of foil-like manganin gauges in planar shock wave experiments. *J. Appl. Phys.* **1980**, *51*, 3702–3705. [CrossRef]
13. Rosenberg, Z.; Brar, N.S. The influence of the elasto-plastic properties of piezoresistance gauges on their loading-unloading characteristics as lateral shock stress transducers. *J. Appl. Phys.* **1995**, *77*, 1443–1448. [CrossRef]
14. Bourne, N.K.; Rosenberg, Z. Simultaneous manganin gauge and visar measurements of shock wave profiles. *Proc. Am. Inst. Phys.* **1980**, *429*, 849–852.
15. Bruun, E. Constant-bandwidth current mode operational amplifier. *Electron. Lett.* **1991**, *27*, 1673. [CrossRef]
16. Romero, J.R.G.; Garcia-Torales, G.; Rosas, G.G.; Ocaña, J.L.; Flores, J.L. Characterization of mechanical shock waves in aluminum 6061-T6 using a high power laser pulse. *Infrared Remote Sens. Instrum. XXIV* **2016**, *9973*, 99730. [CrossRef]
17. Peyre, P.; Fabbro, R. Laser shock processing: A review of the physics and applications. *Opt. Quantum Elect.* **1995**, *27*, 1213–1229.
18. Zhang, Z.; Nian, Q.; Doumanidis, C.C.; Liao, Y. First-principles modeling of laser-matter interaction and plasma dynamics in nanosecond pulsed laser shock processing. *J. Appl. Phys.* **2018**, *123*, 054901. [CrossRef]
19. Scholl, M.S. Spatial and temporal effects due to target irradiation: A study. *Appl. Opt.* **1982**, *21*, 1615–1620. [CrossRef] [PubMed]
20. Morales, M.; Porro, J.; Blasco, M.; Molpeceres, C.; Ocaña, J.L. Numerical simulation of plasma dynamics in laser shock processing experiments. *Appl. Surf. Sci.* **2009**, *255*, 5181–5185. [CrossRef]
21. Trdan, U.; Porro, J.A.; Ocaña, J.L.; Grum, J. Laser shock peening without absorbent coating (LSPwC) effect on 3D surface topography and mechanical properties of 6082-T651 Al alloy. *Surf. Coat. Technol.* **2012**, *208*, 109–116. [CrossRef]
22. Rubio-González, C.; Gomez-Rosas, G.; Ocaña, J.; Molpeceres, C.; Banderas, A.; Porro, J.; Morales, M. Effect of an absorbent overlay on the residual stress field induced by laser shock processing on aluminum samples. *Appl. Surf. Sci.* **2006**, *252*, 6201–6205. [CrossRef]
23. Trdan, U.; Skarba, M.; Grum, J. Laser shock peening effect on the dislocation transitions and grain refinement of Al–Mg–Si alloy. *Mater. Charact.* **2014**, *97*, 57–68. [CrossRef]
24. Scholl, M.S. Measured spatial properties of the cw Nd:YAG laser beam. *Appl. Opt.* **1980**, *19*, 3655–3659. [CrossRef] [PubMed]
25. Gonzalez-Romero, R.; Garcia-Torales, G.; Gomez Rosas, G.; Strojnik, M. Spatial Dependence of a Laser-Induced Shock Wave Mitigator Matrix: A Numerical Study. In Proceedings of the SPIE Optical Engineering Applications, San Diego Convention Center, San Diego, CA, USA, 20 August 2020; Strojnik, M., Ed.; SPIE: San Diego, CA, USA, 2020.
26. Duan, Z.; Liu, Y.; Pi, A.; Huang, F. Foil-like manganin gauges for dynamic high pressure measurements. *Meas. Sci. Technol.* **2011**, *22*, 075206. [CrossRef]
27. Hu, H.; Liu, T.; Zhai, H. Comparison of femtosecond laser ablation of aluminum in water and in air by time-resolved optical diagnosis. *Opt. Express* **2015**, *23*, 628–635. [CrossRef] [PubMed]

photonics

MDPI

Communication

In-Situ Depth Measurement of Laser Micromachining

Xiaoming Chen [1], Ying Xu [1], Nan-Kuang Chen [2], Shannon Shy [3] and Hsiang-Chen Chui [1,3,*]

1 School of Optoelectronic Engineering and Instrumentation Science, Dalian University of Technology, Dalian 116024, China; chen_xm@dlut.edu.cn (X.C.); 32042038@mail.dlut.edu.cn (Y.X.)
2 School of Physics Sciences and Information Technology, Liaocheng University, Liaocheng 252000, China; nankuang@gmail.com
3 Department of Photonics, National Cheng Kung University, Tainan 70101, Taiwan; ss710892@gmail.com
* Correspondence: hcchui@dlut.edu.cn

Abstract: Precision laser micromachining plays an important role in the biomedical, electronics, and material processing industries. During laser drilling, precision depth detection with micrometer-level resolution is required, particularly with blind-hole or heterogeneous structures. We present an optical detection system utilizing an optical confocal structure, experimentally confirmed to achieve a >95% accuracy for micron-diameter holes that are tens-of-microns deep. This system can be easily integrated into commercial laser micromachining processes, and can be employed in laser drilling and three-dimensional active-feedback laser printing.

Keywords: laser machining; depth measurement; confocal structure

1. Introduction

Precision laser micromachining plays a key role in the biomedical, electronics, and material processing industries, providing high-precision control of drilling depth for blind holes and heterogeneous structures. Q-switched lasers, which are also known as giant pulse lasers, can deliver giant pulses for use in laser machining, with their ablation rate during drilling decreasing with hole depth [1]. The drilling rate is highly nonlinear with depth and machining time, due to a number of effects involving the laser drilling process, including laser defocusing, the production of laser-induced plasmas, and varying material specifications. Several inline and repeatable methods designed for real-time monitoring or in-situ measurements of the laser ablation depth have been reported but continue to have limitations. Currently, the standard industrial process is to adopt a trial-and-error approach to determine laser machining curves for each sample, which outline the drilling depth over time under specific laser conditions. However, during laser drilling, machining conditions may be affected by unpredicted circumstances, such as laser power decay, clearance of the focusing system, optical misalignments resulting from mechanical vibrations, and human action. High throughput and quality assurance are required during offline, high-sampling-rate inspections, and are, therefore, costly. Indeed, many techniques for monitoring of laser ablation depth in real-time have been reported.

Optodynamical methods for monitoring laser micro-drilling have been proposed and developed by Yeack et al. [2]. One can verify that laser drilling occurs by simultaneously detecting the laser-induced ultrasonic waves in a workpiece and in its surrounding air caused by ablation [3]. Such methods can directly measure the hole depth using piezo electric pressure sensors [4] or a laser beam deflection probe (LBDP) [5] that measures the time-of-flight of the shock resulting from laser ablation. Use of a LBDP is implemented experimentally by means of a digital micrographic system that enables the acquisition of images of the plasma plume and the hole cross-section during a drilling sequence. Optodynamic techniques have been applied to determining ablation depth in situations where the time-of-flight of the acoustic shock wave originating from a pulse can be detected using piezoelectric or interferometric techniques as it exits the rear of a material.

This approach is easy to set up and can be integrated with commercial laser machining systems. However, one challenge in using this method is that workpieces may be heterogeneously structured. A technique that uses open piezoelectric acoustic sensors placed above the workpiece has been developed, but the characteristics of the acoustic pulse are dependent on the material properties [5] making it unsuitable for industrial situations where differing or non-homogeneous materials are common. Another approach to the real-time monitoring of laser processing uses the current induced by an electric field applied to a laser-produced plasma [6]. The resulting design for a new electrode circuit has allowed for an increased probing distance between the electrodes and the processed sample surface, thereby avoiding the direct contact of the electrodes with the plasmas and reducing sputtering. Papazoglou and coworkers [7] demonstrated the depth-resolved analysis of multi-layer structures is carried out by employing laser-induced breakdown spectroscopy (LIBS) and white light interferometry, which enables accurate in-situ depth monitoring. On-line LIBS determination of magnesium coating thickness [8] was performed by Ruiz et al. Several optical distance-measurement techniques have been proposed and implemented, with nanometer precision and the multi-millimeter depth ranges required in laser micro-hole drilling. Direct distance measurement techniques utilizing an autofocusing probe using two cylindrical lenses for profile measurement [9] or pure interferometric techniques [10,11] have been used to measure the distances required with great accuracy, and could be suitable for ablation depth measurement. Direct interferometric techniques have been used to measure micron and sub-micron distances in several applications [9–11], but such techniques rely on high-intensity, directly-reflected light to discern the interference fringes. This limitation has been overcome in optical coherent tomography [12], where a coherent confocal imaging system is used to determine distance [13]. However, this method lacks simplicity and can be difficult to integrate into an industrial setting. All optical techniques are of value in their particular applications, but are inadequately flexible when applied to detecting laser-generated micron-scale hole depths in difficult geometries.

In this work, we present an approach for the inline, precise monitoring of laser ablation depth as applied to blind-hole laser drilling for single and multilayer bulk heterogeneous structures, and for different materials, an alloy and a silicon wafer. The detection method was based on the reflection of the probe beams from the bottom surfaces of the drilled holes. Using theoretical modeling and practical use cases, this method can be implemented to control error to within 5% for 100-μm hole depths.

2. Materials and Methods

The principle of confocal structure was described by Marvin Minsky in 1957 [14] and has been applied to many sophisticated laser-scanning systems designed for a range of biomedical applications. Major improvements to the instrumentation photon efficiency, coupled with the development of novel fluorescent reporters, have enabled the multi-dimensional imaging of living cells and tissues [15]. Confocal microscopy uses point illumination and a pinhole in an optically conjugate plane in front of the detector to filter out defocused image information. Only light emanating from the focal plane is detected, allowing image quality to be far superior to that of wide-field fluorescence images. More-over, the thickness of the focal plane is inversely proportional to the square of the numerical aperture of the objective lens used, and is dependent on the optical properties of the specimen and ambient index of refraction [12]. This model of confocal microscopy applies to the confocal structure of laser ablation depth detection. A confocal microscope only receives the signal reflected back from the focal plane of the objective in a specimen, a property of confocal systems that is implemented here. By scanning the pinhole, the position of maximum signal intensity can be found that corresponds to the depth of a hole′s bottom surface being drilled during laser machining.

The method proposed here uses a probe laser beam to illuminate the workpiece with a confocal structure. The experimental scheme is shown in Figure 1. The confocal optical system was integrated with a commercial laser machining station. The machining laser was a 1064-nm Q-switched Nd:YVO_4 laser. The probe laser was a continuous wave 50-mW 532-nm frequency-doubling Nd:YVO_4 laser. Its beam size was about 1 mm without focusing lens. The pulse repetition frequency was tunable from 1 to 120 kHz. A set of half-wave plates (HWP) and a polarizer (PL) were placed after the machining and probe lasers in order to allow for output power adjustment. A 45° dichroic mirror (DM) was employed to combine the beams from the machining and probe lasers.

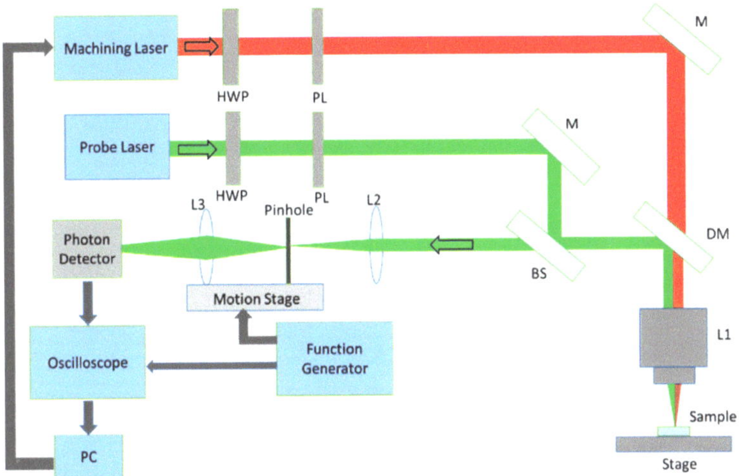

Figure 1. Schematic optical layout for real-time depth monitoring. HWP: Half-Wave Plate; PL: Polarizer; DM: Dichroic Mirror; BS: Beam Splitter; M: Mirror; and PC: Personal Computer.

The probe beam was reflected by a 50/50 beam splitter (BS) and a DM, focused on the workpiece by Lens 1 (L1), and then reflected back from the bottom surface of the ablated area of the workpiece where the beam was gathered by the same focusing lens. The quasi-collimated gathered light was then reflected by a DM, passed through a BS, and focused on a 10-μm-diameter pinhole by another focusing lens (L2). The diffracted light after the pinhole was focused on a photo detector (PD) by a focusing lens (L3). The signals from the PD were sent to the oscilloscope connected to a computer. The pinhole and focusing lens (L3) were arranged on a linear motion stage (New Focus 9066, Newport, IR, USA) driven by an iPico module (New Focus 8763, Newport, IR, USA). The simultaneous search for the maximum intensity position on the PD while the line motional stage moved was driven by a function generator with triangle waveform. The laser drilling process on the samples typically induced the debris, nanosized dust, and even plasma. They may cause the fluctuation along the optical path and result in the measurement noises. So, we performed the measurements after the laser drilling processes. Each position scan of a linear motion stage costs 1.2 s. Each measurement process was performed after the laser drilling process. And the 0.5-s interval between them. The durations of data acquisitions and the depth calculations of the drilled hole was less than 0.1 s. A complete drilling and measurement cost about 3 s. Therefore, the operation time was still long for mass productions. Higher measurement accuracy for a pulse-to-pulse operation within 1 s was still studied.

Inserting a BS into the optical path of a machining laser can result in defocusing. This defocusing results in z-position shift of the focal point, and can be solved by finding a new z-axis distance. In short (ns) and ultrashort pulse (ps/fs) laser machining, pulse shape deformation may be induced by optical elements in the optical path [16]. Here, the BS thickness was 2 mm and the pulse duration of the machining laser was 10 ns.

After checking the pulse shape by a PD with a gigahertz response frequency, pulse shape deformation was monitored.

To increase the position sensitivity, the focal lengths of Lenses 1 and 2 were chosen. We analyzed a theoretical model by numerically simulating two cases, and then compared the experimental data with the model.

3. Theoretical Model Using Numerical Simulation

A schematic of the test platform is shown in Figure 2a. The optical layout of Figure 1 was simplified as Figure 2a. The focal length of the machining laser focusing lens (L1) in the two cases was set to 50 mm. For the confocal lens (L2), the focal length was set to 50 mm in Case 1, and 100 mm in Case 2. The distance (d) between lenses L1 and L2 was the sum of their focal lengths. The results of the simulation are presented in Figure 2b. In Case 1, the travel distance of the focusing lens shifted 1 mm, and the intensity peak in the confocal setup shifted 1 mm, yielding a lateral sensitivity ratio of 1:1 over 1 mm of travel. In Case 2, the travel distance of the focusing lens shifted 1 mm, and the intensity peak in the confocal setup shifted 4 mm, yielding a lateral sensitivity ratio of 1:4 over 1 mm of travel. Some degradation in the resolution of the system was observed due to mechanical vibration and imperfect optical alignment. The simulation works were performed with Matlab.

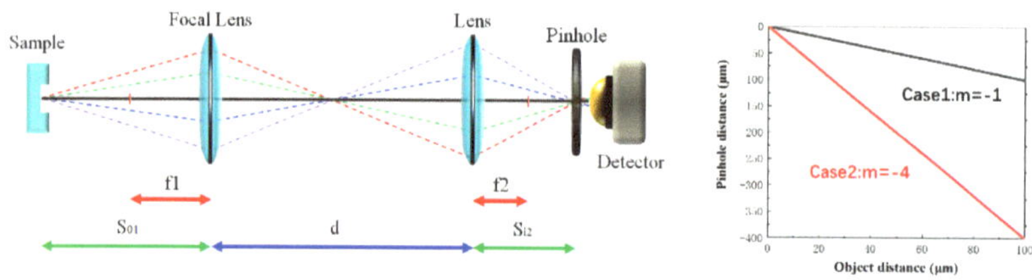

(a) (b)

Figure 2. (**a**) Illustration of simplified confocal depth detection system; and (**b**) simulation results of the lateral sensitivity gain of 1:1 (blue line) and 1:4 (red line) over 100 μm of travel, respectively. (Case 1: f_1 = 50 mm, f_2 = 50 mm; Case 2: f_1 = 50 mm, f_2 = 100 mm).

4. Results and Discussion

4.1. Confocal System Tested Using Polished Stainless Steel

The optical system was configured to utilize a probe laser as previously described, and parameters were chosen as discussed in the numerical simulation section. A 10-μm pinhole was utilized for the confocal element, and a confocal lens with a focal length of 100 mm and a suitable collecting lens were chosen to focus the light passing through the pinhole onto the PD.

A piece of polished stainless steel was first affixed to the laser workstation as a sample workpiece. Initially, the probe laser beam was aligned collinearly with the machining laser beam. The vertical stage was fine-tuned through a portion of its range to simulate the movement of the surface of the machined bottom of the drilled hole. The relation of the object distance and the pinhole distance using acquired data is plotted in Figure 3a. The sample was measured at different vertical positions in 20 μm increments from 0 to 120 μm. The data obtained while scanning the object distance at varied vertical positions using Cases 1 and 2 are shown in Figure 3b,c, respectively. In Figure 3a, the root-mean-square (RMS) uncertainty of curve fitting for Case 1 was calculated as 0.046 μm. And, the RMS uncertainty of curve fitting for Case 2 was 0.025 μm. These experimental results confirm that, as indicated by the theoretical and numerical studies, the confocal mechanism for micron-precision hole depth detection effectively tracked the change in relative distance of the workpiece. In particular, as the workpiece moved away from the focus of the focusing lens, the signal intensity and signal-to-noise ratio decreased.

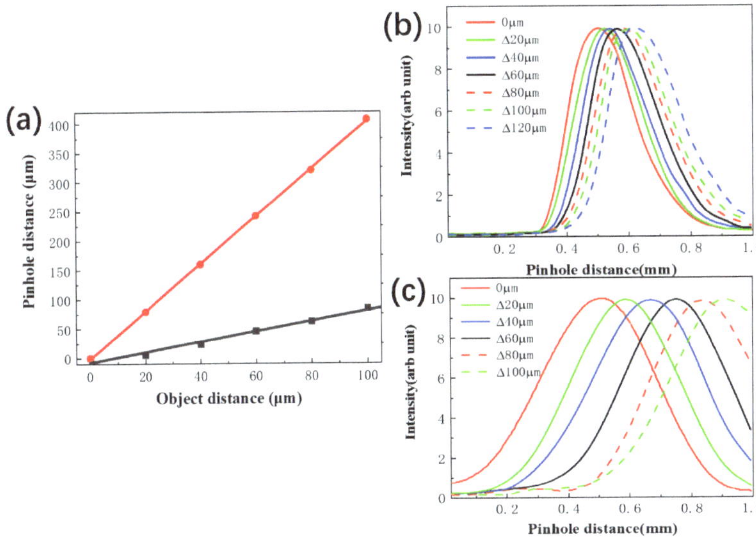

Figure 3. (**a**) Relation of the object distance and a pinhole distance, (**b**) data from Case 1, and (**c**) data from Case 2.

4.2. The Designed Experiments for Laser Drilling in Aluminum Alloy

A series of experiments was designed to showcase the capability of the system. After the alignment of the optics was checked, a small piece of aluminum (Al) alloy was cut and prepared, then affixed to the stage. Then, the Al alloy workpiece was drilled by 20–60 pulses by focusing a 1064-nm pulsed laser with a 0.5-J pulse energy onto the workpiece. The pinhole was then scanned over its entire range of travel, and data were collected with the depth-monitoring system. After the sample was drilled and measured, a three-dimensional (3D) image map of the sample was obtained using a 3D laser microscope (Keyence VK-9700, Tokyo, Japan).

The surface of the Al alloy was drilled by 20 pulses with 0.5-J pulse energies. Figure 4a shows a 3D image map from the laser microscope, with a cross-section included as Figure 4b. Figure 4c,d show the scan data using Cases 1 and 2, respectively. The hole depth was measured to be 34.04 μm using the 3D laser microscope. In Case 1, the hole depth was measured to be 35 μm with the confocal optical design, with an error of ~2.94%. In Case 2, the position of the maximum intensity measured was 136 μm, so assuming the numerical simulation of the 4:1 ratio of transverse position, the hole was calculated to be 33.75 μm deep with an error of ~0.85%. At different drilling depths, the number of machining laser pulses was increased to 40 and 60 pulses. These experimental data are shown in Figures 5 and 6, and summarized in Table 1. The experimental processes were performed more than three times, and the error was generated from the standard derivation of data sets. The more than 10 % measurement variations occurred and when the hole depth was larger than 170 μm.

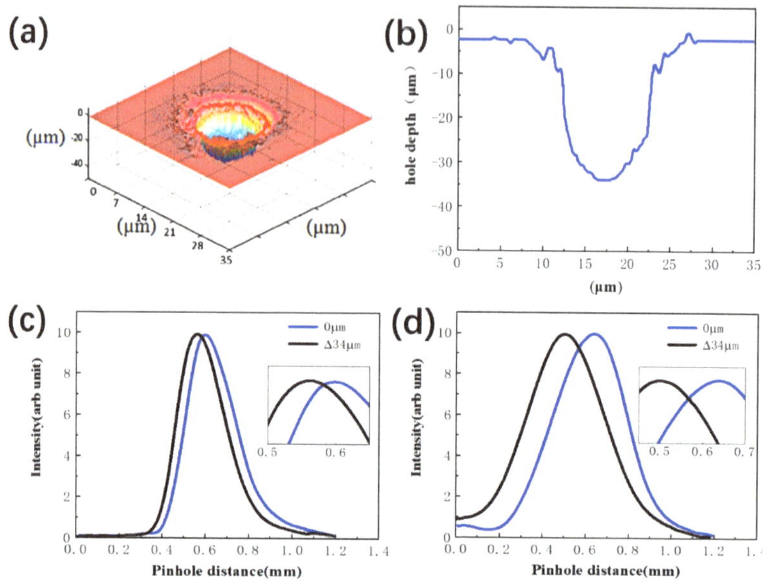

Figure 4. Image of the drilled hole after laser drilling with 20 pulses. (**a**) 3D confocal method; (**b**) cross-section; and (**c**,**d**) data from Cases 1 and 2, respectively.

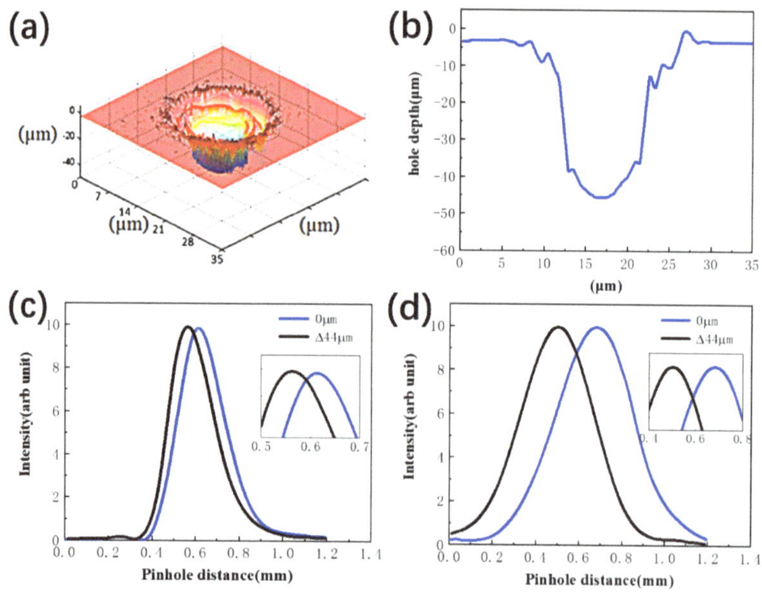

Figure 5. Image of the drilled hole after laser drilling with 40 pulses. (**a**) 3D confocal method; (**b**) cross-section; and (**c**,**d**) data from Cases 1 and 2, respectively.

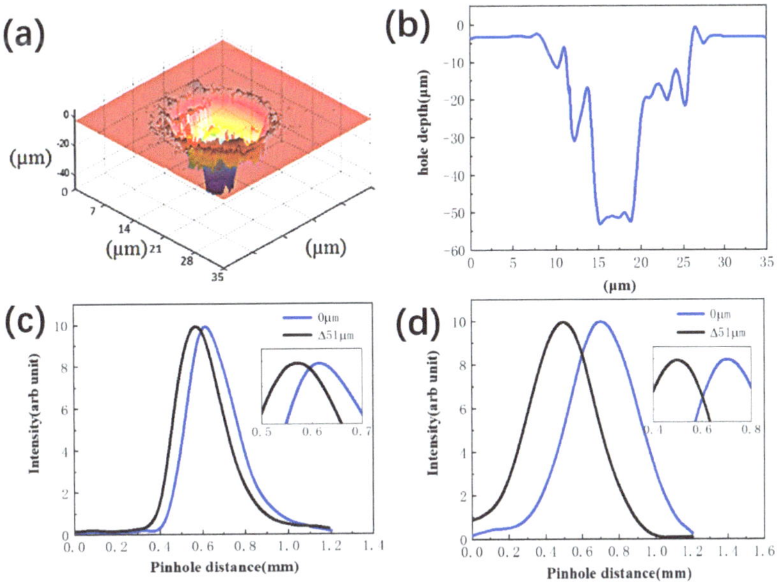

Figure 6. Image of the drilled hole after laser drilling for 60 pulses. (**a**) 3D confocal method; (**b**) cross-section; and (**c**,**d**) data from Cases 1 and 2, respectively.

Table 1. Summary of experimental data.

Number of Laser Pulses	Hole Depth and Error (Unit: μm (%))		
	3D Laser Microscope	Case 1	Case 2
20	34.04	35 (2.94%)	33.75 (0.85%)
40	44.81	45 (0.44%)	45.5 (1.56%)
60	51.15	50.4 (1.37%)	53 (3.71%)

5. Conclusions

This study performed theoretical simulations and experiments pertaining to the development of an all-optical real-time laser drilling depth detection method. The approach is based on a confocal optical structure, achieves a precision of <1 μm and an inaccuracy of <5% for ~100-μm-deep holes. This precision and accuracy are sufficient for real-time machining depth monitoring during laser micro-hole drilling, and the system is sufficiently robust for use in industrial and research applications. Furthermore, this method is independent of the wavelengths of the machining lasers, and can be integrated with commercial laser-machining systems in general. A reliable control process can be easily implemented for automatic control of machining depth in applications that are sensitive to the required depth of blind holes, eliminating the need for expensive back-wall protection, and highly-skilled and experienced operators. A >100-μm-deep detections can be realized with a low-divergence-angle beam on the targeted sample surfaces. This study is still on-going.

Author Contributions: Conceptualization, X.C. and H.-C.C.; Data Measurement, Y.X. and S.S.; Data Analysis, Y.X. and N.-K.C.; Writing and Supervision, H.-C.C. All authors have read and agreed to the published version of the manuscript.

Funding: Ministry of Science and Technology, Taiwan (MOST 107-2622-E-006-014-CC3), and by the Fundamental Research Funds for the Central Universities, China (DUT18RC(3)047 and DUT20RC(5)028).

Institutional Review Board Statement: Not applicable.

Informed Consent Statement: Not applicable.

Data Availability Statement: The data presented in this study are available in the article.

Conflicts of Interest: The authors declare no conflict of interest.

References

1. Petkovsek, R.; Panjan, I.; Babnik, A.; Mozina, J. Optodynamic study of multiple pulses micro drilling. *Ultrasonics* **2006**, *44*, E1191–E1194. [CrossRef]
2. Yeack, C.E.; Melcher, R.L.; Klauser, H.E. Transient photo-acoustic monitoring of pulsed laser drilling. *Appl. Phys. Lett.* **1982**, *41*, 1043–1044. [CrossRef]
3. Strgar, S.; Mozina, J. An optodynamic determination of the depth of laser-drilled holes by the simultaneous detection of ultrasonic waves in the air and in the workpiece. *Ultrasonics* **2002**, *40*, 791–795. [CrossRef]
4. Stafe, M.; Negutu, C.; Popescu, I.M. Real-time determination and controlof the laser-drilled holes depth. *Shock Waves* **2005**, *14*, 123–126. [CrossRef]
5. Petkovsek, R.; Babnik, A.; Diaci, J.; Mozina, J. Optodynamic monitoring of laser micro-drilling of glass by using a laser probe. *Appl. Phys. A-Mater. Sci. Process.* **2008**, *93*, 141–145. [CrossRef]
6. Idris, N.; Madjid, S.N.; Ramli, M.; Kurniawan, K.H.; Lee, Y.I.; Kagawa, K. Monitoring of laser processing using induced current under applied electric field on laser produced-plasma. *J. Mater. Process. Technol.* **2009**, *209*, 3009–3021. [CrossRef]
7. Papazoglou, D.G.; Papadakis, V.; Anglos, D. In situ interferometric depth and topography monitoring in LIBS elemental profiling of multi-layer structures. *J. Anal. At. Spectrom.* **2004**, *19*, 483–488. [CrossRef]
8. Ruiz, J.; González, A.; Cabalín, L.M.; Laserna, J.J. On-Line Laser-Induced Breakdown Spectroscopy Determination of Magnesium Coating Thickness on Electrolytically Galvanized Steel in Motion. *Appl. Spectrosc.* **2010**, *64*, 1342–1349. [CrossRef]
9. Rhee, H.G.; Kim, D.I.; Lee, Y.W. Realization and performance evaluation of high speed autofocusing for direct laser lithography. *Rev. Sci. Instrum.* **2009**, *80*, 073103. [CrossRef]
10. Lading, L.; DamHansen, C.; Rasmussen, E. Surface light scattering: Integrated technology and signal processing. *Appl. Opt.* **1997**, *36*, 7593–7600. [CrossRef]
11. Hofstetter, D.; Zappe, H.P.; Dandliker, R. Optical displacement measurement with GaAs/AlCaAs-based monolithically integrated Michelson interferometers. *J. Lightwave Technol.* **1997**, *15*, 663–670. [CrossRef]
12. Webb, R.H. Confocal optical microscopy. *Rep. Prog. Phys.* **1996**, *59*, 427–471. [CrossRef]
13. Arons, E.; Leith, E. Coherence confocal-imaging system for enhanced depth discrimination in transmitted light. *Appl. Opt.* **1996**, *35*, 2499–2506. [CrossRef] [PubMed]
14. Minsky, M. Memoir on inventing the confocal scanning microscope. *Scanning* **1988**, *10*, 128–138. [CrossRef]
15. Paddock, S. Over the rainbow: 25 years of confocal imaging. *Biotechniques* **2008**, *44*, 643–648. [CrossRef]
16. Tan, B. Deep micro hole drilling in a silicon substrate using multi-bursts of nanosecond UV laser pulses. *J. Micromech. Microeng.* **2006**, *16*, 109–112. [CrossRef]

photonics

Article

Influence of Spatio-Temporal Couplings on Focused Optical Vortices

Anda-Maria Talposi [1,2], Vicentiu Iancu [1,2] and Daniel Ursescu [1,2,*]

[1] ELI-NP, Horia Hulubei National Institute for Physics and Nuclear Engineering, 30 Reactorului Street, 077125 Magurele, Romania; maria.talposi@eli-np.ro (A.-M.T.); vicentiu.iancu@eli-np.ro (V.I.)
[2] Faculty of Physics, University of Bucharest, 405 Atomistilor Street, 077125 Magurele, Romania
* Correspondence: daniel.ursescu@eli-np.ro

Abstract: Ultra-intense laser pulses with helical phases are of interest in laser-driven charged particle acceleration and related experiments with extreme light. However, such optical vortices can be affected by the presence of residual spatial-temporal couplings. Their field distributions after propagating in free-space and in the focal plane of an ideal focusing mirror were assessed through numerical modeling, based on the Gaussian decomposition method for a 25 fs pulse with a Super-gaussian spatial profile. The wash-out of the central hole in the doughnut-shaped profile in the focal plane corresponds to the rotation of the phase discontinuity.

Keywords: high-power lasers; ultra-short pulses; helical phase; optical vortex; spatio-temporal couplings; laser pulse propagation; Gauss decomposition

1. Introduction

Femtosecond laser systems have opened new frontiers in the study of matter at an ultrafast timescale, using broad-spectral-bandwidth pulses. The chirped pulse amplification (CPA) technique [1], combined with broad-gain-bandwidth optical parametric chirped pulse amplification (OPCPA) [2], made it possible to observe peak powers in excess of 10^{16} W, as reported for the high-power laser system (HPLS) at the Extreme Light Infrastructure-Nuclear Physics (ELI-NP) facility [3].

Ever since the concept of optical vortices (OVs) was proposed in [4,5] and then observed experimentally in [6,7], the continuous interest and developments in his area have spawned a broad range of potential applications, including their use high-power lasers. The idea that orbital angular momentum (OAM) can exist within OVs, and a method of generating it, was first suggested by Allen et al. [8], thus providing a new way to study the effects of the connection between quantum and classical optics (paraxial beams). Studies that involve OVs range from research on the use of optical tweezers for particle trapping and manipulation [9], quantum applications [10–13], biomedical applications [14], super-high-resolution imaging [15–17], optical communications [18], and ultraviolet and X-ray light [19,20].

OVs can be generally defined as a stream of photons propagating with a singularity in the phase field. The helical phase associated with the light beam corresponds to a spiral rotation of the wavefront along the direction of propagation.

A unique property of OVs resides in the fact that the azimuthal gradient associated with the propagating helical phase, $\exp(il\vartheta)$, is responsible for the OAM, where ϑ is the angle cylindrical coordinate and l is an integer called topological charge. Therefore, the OAM component, directed along the propagation axis, is known as helicity [21]. This means that the phase exhibits a turn/revolution of 2π radians around the dislocation axis over an interval of l wavelengths. The amplitude of the electric field of the light wavefronts becomes zero in the dislocation center.

The light beams with OAM are represented in terms of LG_p^l (Laguerre–Gaussian) modes, where l gives the number of intertwined helices (azimuthal indices), and p, the number of radial modes. LG_0^1 beams, also known as "doughnut" beams, can be set apart from the LG_0^0 modes (Gaussian beams) due to the zero-intensity dark spot at the center of their beam profile and the helical wavefront with singularity in the center, generating a screw-like dislocation in the electric field structure.

It has been shown in theory that helical beams can be used in the direct laser acceleration of electrons, and also that the OAM associated with helical beams can be partially transferred to electrons [22]. Experimentally, it was demonstrated in [23], that upon generating OV beams in a PW laser system, the energy of ions accelerated by a helical laser beam was lower compared to the laser-driven ions with Gaussian focal spots, but the gain in energy was higher for the same initial laser pulse energy.

Light beams with OAM can be produced in laboratory conditions using various methods, including the control of laser modes in the cavity, diffractive optical elements, lenses, spiral phase plates, helical phase plate mirrors and spatial light modulators [24]. For high power laser systems, spiral phase plates are available for the generation of ultra-intense OV. As a consequence, several theoretical and experimental investigations have been carried out.

Experiments that require intense laser fields with a helical shape can be compromised if the helical focus is deformed. Therefore, studying and understanding possible sources of distortion will allow the optimization of the focal spot and provide useful input information for the analysis of the experimental data.

In high-power laser systems, the beam diameters, as well as the bandwidths, are very large (hundreds of millimeters and tens of nanometers, respectively). This is why the variation of the temporal properties across the spatial beam profile is not negligible and may cause detrimental distortions of the field on target. These distortions are of the spatio-temporal type and are generated by so-called spatio-temporal couplings (STC). Even careful design and alignment procedures cannot ensure a perfectly smooth spatio-temporal field. Residual STC can originate from dispersive optical components in the beam path, from small defects and even from a minimal misalignment of the temporal compressor [25,26].

The effects of STC create specific patterns in the focus region of helical high power laser fields. In order to assess their impact, we present here a theoretical model, based on the Gaussian decomposition method, of the free-space propagation of pulsed optical vortices in the spatio-temporal domain. The propagation code simulates the cases in which a high-power laser field with a helical spatial profile, with or without residual STC, propagates inside the transport system under vacuum, towards the target place.

The paper is organized as follows. In Section 2, we present the theoretical framework of the numerical calculations, the characteristic properties of optical vortices and the characteristics of the input laser field. In Section 3, the results pertaining to helical laser fields, with different spatio-temporal distortions, such as spatial chirp, angular dispersion and pulse front tilt, are shown. Section 4 concentrates on the specific physics aspects of the ultrashort optical vortex propagation phenomena. Section 5 summarizes the conclusions we drew from this analysis.

2. Description of the Method

The propagation of laser fields has been a widespread concern, at first for monochromatic (narrowband) beams and then for pulsed, broadband lasers. Although the monochromatic cases can be approached using ray tracing, the beam propagation method (BPM), diffraction integrals based on Huygens' principle or Fourier optics, the propagation of pulsed-beams is more complex. The evolution in space and time of pulsed laser fields can be calculated as a superposition of monochromatic waves, applying the Fourier temporal transformation, but one can also use more rigorous approaches such as solving the Maxwell equations via the finite-difference time-domain (FDTD) technique.

More recently, a Gaussian decomposition method has been used to determine the propagation of pulsed beams. The method consists in the decomposition of the laser field as a superposition of Gaussian beams, which are then individually propagated to the point of interest, and then the reconstructed field in the region of interest is obtained as the superposition of the propagated Gaussian fields. Although this method was initially used for monochromatic beams [27,28], it can also be extended to time-limited waves [29,30].

The results in this paper were obtained using the propagation method described in greater detail in [30]. First, the amplitude of the spatial field is defined: here, a Supergaussian profile of order $n = 6$ was used, with the following distribution on x and y:

$$E_{s,S}(x, y, z = 0) = A_{0S} \times \exp\left[-\left(\frac{\sqrt{x^2 + y^2}}{w_{0S}}\right)^n\right] + N. \tag{1}$$

A_{0S} is the amplitude of the Supergaussian function, which was set to $A_{0S} = 1$ in this work. w_{0S} is the width (half-diameter) and was set to $w_{0S} = 55/2$ mm, to have similar parameters as those of the 100 TW beamline from the HPLS at ELI-NP in Romania [3]. The initial phase is considered to be flat (zero). N is a small random noise of amplitude 0.002. The Supergaussian shape of the beam profile is also relevant because it is often used in CPA laser systems, as it provides optimal energy extraction from the laser amplifiers. Therefore, an LG mode cannot be generated in high-power lasers, but a "modified LG mode" can be obtained using specific spiral optical elements [31].

The spatial distribution of the electric field in the xy plane is then decomposed into many Gaussian terms, using the fitting algorithm from Wolfram Mathematica [32]. Here, 121 terms are distributed on an 11×11 rectangular grid. The center $X_{0i,j}$ and $Y_{0i,j}$ of each Gaussian is allowed to vary during slightly the fitting process. Therefore, the Gaussian parameters of amplitude $E_{0i,j}$, central positions $X_{0i,j}$ and $Y_{0i,j}$, and waist $W_{0i,j}$, are obtained for each element i, j in the decomposition such that

$$E_{s,S}(x, y, z = 0) \approx \sum_{i,j=1}^{11} E_{0i,j} \times \exp\left[-\left(\frac{(x - X_{0i,j})^2 + (y - Y_{0i,j})^2}{W_{0i,j}^2}\right)\right]. \tag{2}$$

Each of these Gaussian terms can be further propagated using the Gaussian beam theory [33], assuming that the waists are placed at $z = 0$. The RMS error for this decomposition was 0.2%, of which approx. 90% was caused by the random noise N in Equation (1). A better accuracy can be obtained by increasing the number of Gaussian terms in the decomposition. However, this number is limited by the fact that the width of each Gaussian should be much larger than the wavelength, to keep the paraxial approximation.

Furthermore, each spatial Gaussian $E_{si,j}(x, y, z)$ becomes time-dependent via a simple multiplication with a narrow-band temporal Gaussian:

$$E_{Gi,j,m}(x, y, z, t) = E_{si,j}(x, y, z) \cdot E_{t,m}(t), \tag{3}$$

the index m refers to element m in the temporal/spectral decomposition.

Therefore, one can neglect the intrinsic spatio-temporal couplings of each Gausslet. However, they can have different parameters from each other, causing their superposition (i.e., the full laser field) to manifest spatio-temporal dependences, as shown here further in Section 3. The temporal Gaussian terms used in Equation (3) are also determined by decomposing an initial broadband pulse using a fitting algorithm that considers both the spectral amplitude and the spectral phase [30].

In this work, the temporal distribution of the initial pulse was considered to be the Gaussian of 25 fs FWHM irradiance at the Fourier limit (FL), centered at $\lambda_0 = 800$ nm. We considered the case in which the pulse has a flat spectral phase, as well as the case in which it is stretched to 4 times its FL pulse duration. Such a temporal distribution was decomposed into 23 pulselets of Gaussian shape, of narrow bandwidths and different central wavelengths [30]. The RMS error of the spectral fit decomposition was 1.2% for the FL pulse.

In this way, the propagation of the initial, full wave in free space or after a focusing mirror can be calculated as a superposition of the spatio-temporal Gausslets from the decomposition. For free-space, one can apply the well-known formula of Gaussian beam propagation:

$$E_{s,G}(x,y,z) = A_s(z) \cdot \exp\left[-\frac{(x-X_0)^2+(y-Y_0)^2}{w^2(z)}\right] \cdot \exp\left[i\frac{k}{2}\frac{(x-X_0)^2+(y-Y_0)^2}{R(z)}\right] \cdot \cdot \exp[-i\varphi(z)] \cdot \exp(ikz). \tag{4}$$

where k is the wavenumber, $A(z)$ the amplitude, $w(z)$ the width at $1/e$ of the maximum electric field at the z position, $R(z)$ is the radius of curvature of the wavefront and $\varphi(z)$ is the Gouy phase [30,33]. Note that the width $w(z)$ is the smallest at the waist, where, conventionally, $z = 0$.

The method used to focus the beam with $f = 1500$ mm focal length optics was to rotate the axis of the Gaussian element towards the focal point and to calculate its new waist $w_{0fi,j,m}$ considering that the initial Supergaussian beam was placed in the front focal plane of the focusing optics.

$$w_{0fi,j,m} = \frac{\lambda_m \cdot f_{i,j}}{\pi W_{0i,j}}, \tag{5}$$

where f_{ij} was determined using the parabola equation:

$$f_{ij} = \sqrt{X_{0,ij}^2 + Y_{0,ij}^2 + \left(\frac{X_{0,ij}^2 + Y_{0,ij}^2}{4f} - f\right)^2} \tag{6}$$

and Equation (5) was found using ABCD matrices [33] for a 2-f system. Note that Equation (5) is different than the one given in [30], but they are approximately the same if $2f_{i,j}\lambda_m \gg \pi W_{0i,j}^2$.

Moreover, to obtain a helical-type spatial phase to the full laser field, the approach was to simulate the reflection of the beam on a "helical" mirror. The mirror would imprint a specific phase distribution upon the beam due to its surface topology, defined by:

$$\Delta z(x,y) = s \cdot \Theta(x,y), \tag{7}$$

where s is the step between the two edges of the helical mirror and $\Theta(x,y)$ is proportional to the cylindrical coordinate $\tan\vartheta = y/x$, with some modifications to include the full $(0, 2\pi)$ domain:

$$\Theta(x,y) = \frac{1}{2} \cdot \left(\frac{\arctan(y/x)}{\pi} - \frac{\operatorname{sign} x}{2}\right). \tag{8}$$

The function $\Theta(x,y)$ that describes the helical wavefront is depicted in Figure 1.

The reflection on a helical mirror described by Equation (7) would imprint a corresponding phase on each i, j, m Gausslet:

$$\frac{2\pi}{\lambda_m} \cdot 2\Delta z(X_{0i,j}, Y_{0i,j}). \tag{9}$$

There is also a shift in the temporal factor of the Gausslet due to the delays caused by the helical mirror. Each individual Gausslet centered initially at $X_{0i,j}$ and $Y_{0i,j}$, $E_{t,m}$ is modified by the terms containing s and therefore becomes dependent on the spatial positions i and j:

$$E_{ti,j,m} = \exp\left[-\left(\frac{t - \frac{z-2\cdot s\cdot\Theta(X_{0i,j},Y_{0i,j})}{c_0}}{\tau_m}\right)^2\right]\cdot\exp(-\mathrm{i}\omega_m t) \tag{10}$$

τ_m and ω_m are the pulse duration and central angular frequency of the m pulselet term in the decomposition. c_0 is the speed of light. Please note the difference between the imaginary number i and the index i.

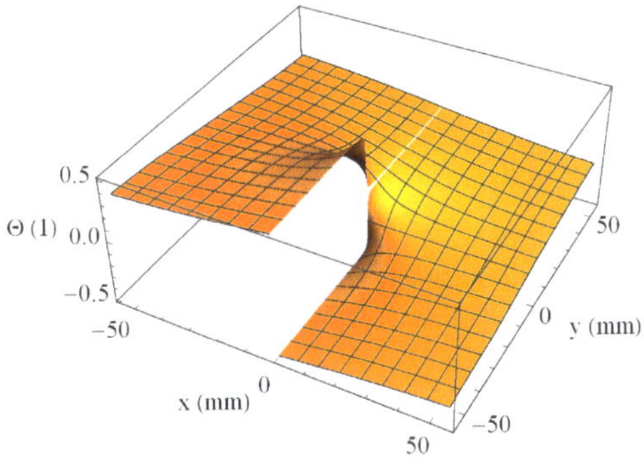

Figure 1. The variation of $\Theta(x, y)$ in Equation (8) with the spatial coordinates x and y, represents the topology of the helical mirror.

3. Results of the Simulations

The propagation code, described previously and detailed in [30], helped to simulate the behavior of the 55 mm Supergaussian laser field, with 25 fs at FL. The full electric field $E(x, y, z, t)$ was reconstructed through the superposition of all the spatio-temporal i, j, m Gausslets.

In the following, several helical cases are presented and discussed: without or with spatio-temporal distortions. Please note that several wavefront distortions were analyzed elsewhere by Ohland and co-authors [31] and will not be discussed here further.

3.1. Ultrashort Laser Fields with Helical Phases

In this subsection, we consider the case of pulses without STC in three configurations: a non-distorted laser field with a helical phase of OAM = 1, at best compression, then with phase jumps corresponding to fractional or higher-order OAM and, finally, with OAM = 1 for pulses chirped in time.

Through this preliminary analysis, we intend to present the ideal case and the impact of small helicity and chirp deviations on the overall field distribution, in the absence of the STC.

3.1.1. Helical Mirror with Matched Step–Wavelength

The case of the perfect helical phase at best compression, OAM = 1 corresponding to $s = \lambda_0/2$ is presented in Figure 2, for the Supergaussian field. The reconstructed wavefront in Figure 2a reproduces the 2π phase jump introduced using Equation (7), corresponding to the theoretical wavefront from Figure 1. The phase values are not relevant on the edges of the plot, because the beam does not cover the full area (the beam size is visible in plot b of Figure 2).

Furthermore, in Figure 2b one can observe the expected "doughnut" shape of the time-integrated irradiance profile associated with the Supergaussian beam, which is visible through the use of detection cameras in the laboratory [34]. The number of Gausslets in the decomposition was 121 (spatial) × 23 (temporal), as previously mentioned.

It is also possible to propagate and reconstruct the complete structure of the field in the focal plane, including the carrier modulation, as presented in Figure 2c. The π phase shift between the upper and lower lobes in the xz plane cut is reproduced, as expected, indicating the spiral field structure in the propagation direction z. Moreover, the channel in the middle is preserved after focusing and the doughnut shape is preserved in the focal plane shown in Figure 2d, as pointed out also in [35]. It is well known that the behavior of the focused fields corresponds to the fields propagating in free space at infinity–known as the far field (FF). Therefore, similar behavior must appear at long propagation distances in free space.

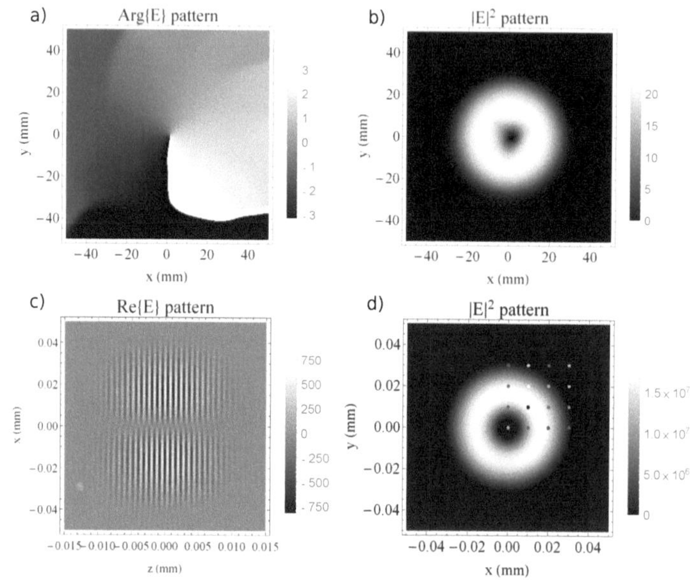

Figure 2. (**a**) The phase profile of the laser field after being reflected by the helical mirror, at $z = 0$. (**b**) The fluence profile ($|E|^2$ time-integrated) of the laser field after being reflected by the helical mirror, at $z = 0$. (**c**) The field profile ($Re\{E\}$) on $x - z$ of the laser field after being reflected by the helical mirror and focused with an $f = 1500$ mm mirror. (**d**) The fluence profile ($|E|^2$ time-integrated) of the laser field after being reflected by the helical mirror and focused with an $f = 1500$ mm mirror. The dots are marking the positions where the temporal profiles are plotted in Figure 3, with the corresponding color.

The pulse temporal shape remains Gaussian in different spatial positions from the doughnut profile. The field varies in intensity, as one can see in Figure 3a–where each temporal profile is represented with the color of the corresponding dot from the fluence profile in Figure 2d. Figure 3b was obtained by normalizing the temporal envelopes in Figure 3a in order to prove that the temporal shape envelope remained the same for all the dots. The shape of the light orange curve corresponds to the one in the center of the doughnut and it is not relevant, as its signal was very low (almost zero compared to the others, i.e., 2200 arb.u. at the peak, or 0.3% of the largest peak, so it cannot be seen in the non-normalized plot). This is also the case for the light green curve, corresponding to the edge of the beam profile.

3.1.2. Helical Mirror with Different Surface Steps

It can be observed sometimes that the doughnut shape is asymmetric, indicated by the fact that one of the lobes is more intense than the other, as in Figure 4a. Simulating the helical laser field introduced above, with different values of the step $s_K = K \cdot s$, showed that variations of a few percentage points in the factor K lead to noticeable lobe amplitude variations.Therefore, not matching the central laser wavelength with the step of the helical mirror causes distortions in the ring shape.

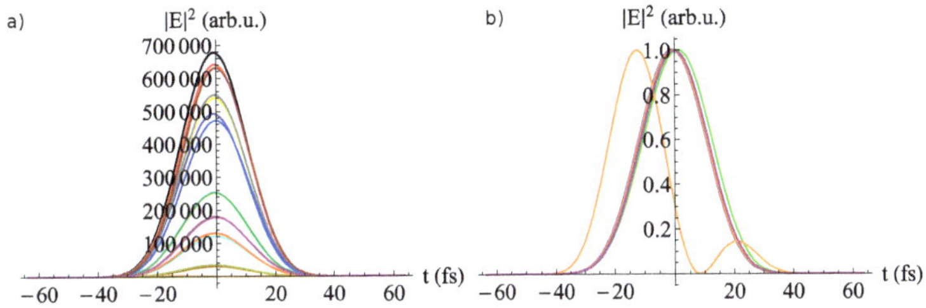

Figure 3. The temporal profile ($|E|^2(t)$) of the laser field after being reflected by the helical mirror and focused with an $f = 1500$ mm mirror. The colors correspond to the positions where the fluence profile is marked in Figure 2d: (**a**) non-normalized and (**b**) normalized. The light orange plot corresponds to $x = y = 0$, where the signal is only an irrelevant, small amount of noise. Normalization was performed so that the maximum of each plot reached 1. The maximum value of the light orange plot appeared close to $t = -20$ fs, whereas at $t = 0$ the signal dropped.

On the other hand, the higher the factor K, the higher the OAM and therefore the beam size increases, as one can see in Figure 4a,b. As an example, the shape for $K = 2$ corresponding to OAM = 2 looks irregular in Figure 4b, this time due to the decomposition of the full beam into a limited number of Gaussian terms (121). The larger the spatial modulations of the beam, the more terms are needed in the decomposition to decrease the reconstruction error. In the following sections, we restrict the analysis to OV with OAM = 1, as this value is the most accessible for practical implementation in high-power laser experiments.

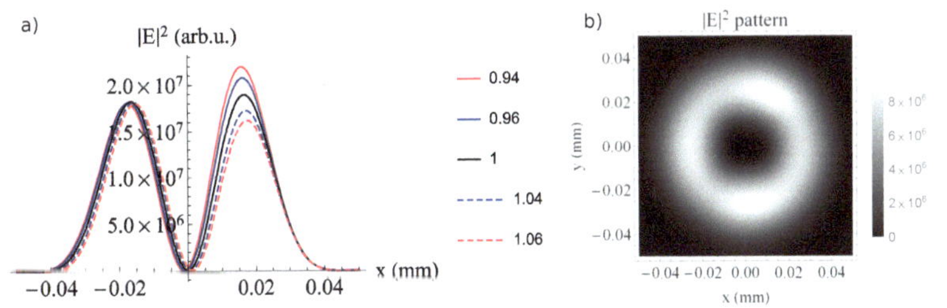

Figure 4. (**a**) The fluence profile ($|E|^2$ time-integrated) on x, when $y = 0$, of the laser field after being reflected by the helical mirror of $s_K = K \cdot s$ and focused with an $f = 1500$ mm mirror. The K value is given in the legend. (**b**) The fluence profile ($|E|^2$ time-integrated) of the laser field after being reflected by the helical mirror with $s_K = 2\frac{\lambda_0}{2}$ and focused with an $f = 1500$ mm mirror.

3.1.3. Chirped Laser Pulse and Helical Spatial Phase

Chirping of the laser field can be achieved by adding dispersion to the FL pulse. Therefore, the spectral amplitude remains the same, but the spectral phase becomes non-flat and the pulse duration increases. In these simulations, the pulse duration was chirped from the pulse width at FL τ_F to $\tau = 4\tau_F$. The spatial phase of this field also becomes helical after reflection on the helical mirror of step s.

The channel in the center, both in the near field (NF), shown in Figure 5a, and in the focus, shown in Figure 5b,c, is similar to the FL case, with the difference that the pulse duration is four times longer. Furthermore, the field amplitude decreases, according to the energy conservation principle. Figure 5a,b demonstrate that the wavefronts exhibited the helical beam characteristic shift between the upper and the lower lobes, indicating that a spiral pattern (and the doughnut) was also preserved. The plots in Figure 5a,b are represented at a smaller scale for z, in order to resolve the wavefront oscillations.

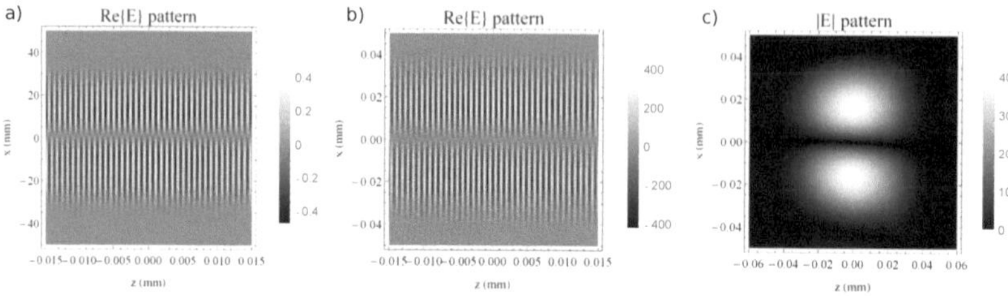

Figure 5. The field profile ($Re\{E\}$ on $x - z$ at $t = 0$ and $y = 0$) of the chirped laser, with duration $\tau = 4\tau_F$ after being reflected by the helical mirror (**a**) in the NF and (**b**) in the focal region of an $f = 1500$ mm mirror (the back focal plane is considered in $z = 0$ for simplicity). (**c**) The corresponding amplitude profile ($|E|$ on $x - z$ at $t = 0$ and $y = 0$) for the focused field shown in plot (**b**).

3.2. *Laser Fields with Helical Phases and Spatio-Temporal Distortions*

In order to model the STC of the initial laser field, specific variations in the parameters of the Gausslets were considered. The helical phase was introduced to these STC fields. The results of the simulations are presented and commented upon in the following.

3.2.1. Spatial Chirp

Spatial chirp (SPC) is the linear variation of the spatial properties for each frequency component in the spectrum: $SPC = \frac{\partial x}{\partial \nu}$ [30,36,37]. Here, we considered the initial Gaussian to be 25 fs FL at $\lambda_0 = 800$ nm, with the temporal pulse shape measured in the spatial center of the beam, at $x = 0$. When there is SPC in the field, then the central frequency of the wave will change according to the SPC variation, meaning that the spectrum will be a different one at each position x: $\nu_{SPC,ijm} = \nu_m + X_{0,ij}/SPC$, where ν_m is the central frequency of the m pulselet in the decomposition. Simulations were performed with SPC on the x axis with the value of $SPC = -1100$ mm/PHz, for a consistent comparison with the non-helical case in [30].

Figure 6 presents the NF, FF and focused distributions of the field in the $y = 0$ plane (upper row) and in the $x = 0$ plane (lower row). Figure 6a,d correspond to the NF profiles on $x - z$ and $y - z$, respectively. In addition to the characteristic fan of the SPC and the channel that appears due to the helicity, a slightly tilted channel in the $x = 0$ plane can be distinguished. In Figure 6b,e the wavefront is still slightly curved, as the divergence of the beams at $3 \cdot z_R$ is not perfectly zero.

As previously mentioned, the behavior of the focused fields corresponds to the fields in the FF. The reconstructed focused field is presented in Figure 6c,f. The wavefront curvature vanishes here, as expected. A gap is apparent on the longitudinal z axis in both FF and focus profiles, which can be associated with the temporal shape, in the $y - z$ profiles. Therefore, there is a spatio-temporal vortex that appears. A comparable behavior of the field was described in [38,39], where the production of spatio-temporal optical vortices was investigated in simulations and in experiments.

Figure 7 presents the time-integrated $|E|^2$ profile in the focus area, that corresponds in practice to an image recorded with a camera sensor. The field in the central part of the doughnut is not going down to zero anymore, as in the case of non-distorted laser pulses. The SPC from the NF generates a tilt in the pulse front in the focus [30] and the time-integrated $|E|^2$ (fluence) profile has a positive value in the center, as shown in Figure 7b. This can be interpreted as a distinctive signature of the presence of SPC in the helical pulse and it appears at relatively small residual SPC values. This sensitivity can be turned into an advantage through the design of a camera-based device that includes a focusing element and a helical phase plate that can detect such small SPC.

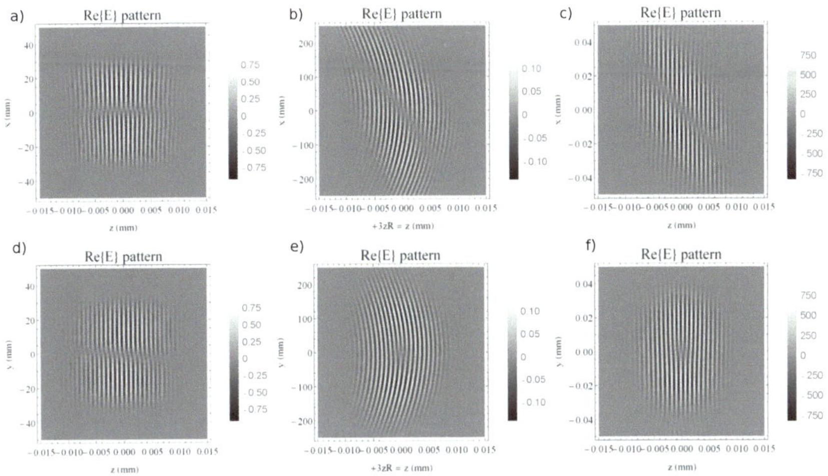

Figure 6. The electric field profile ($Re\{E\}$) of a laser field with $SPC = -1100$ mm/PHz, after being reflected by the helical mirror (**a**) on $x - z$, in the NF ($z = 0$ region), at $t = 0$, (**b**) in the FF at $z = 3 \cdot z_R$ region, at $t = 3 \cdot z_R/c_0$and (**c**) in the focal region after being reflected by the helical mirror and focused by means of an $f = 1500$ mm mirror (at $t = 0$). Similarly, $Re\{E\}$ of the same laser field, but on $y - z$ (**d**) in the NF ($z = 0$ region), (**e**) in the FF at $z = 3 \cdot z_R$ region and (**f**) in the focal region. For simplicity, the z coordinate at $z = 3 \cdot z_R$ and at focal plane was translated to $z = 0$.

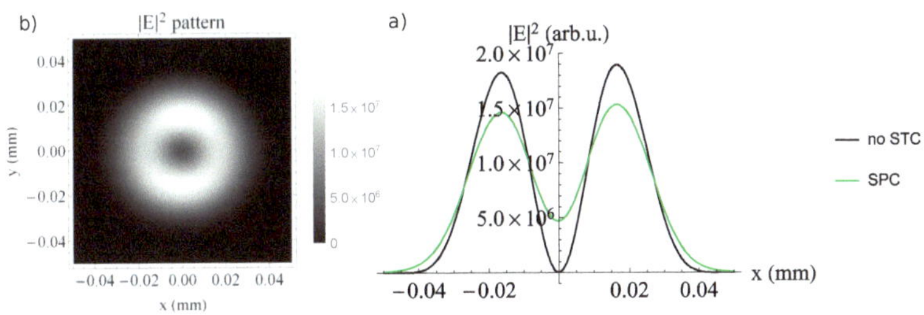

Figure 7. (**a**) The fluence profile ($|E|^2$ time integrated) of a laser field with $SPC = -1100$ mm/PHz, in the focal region after being reflected by the helical mirror and focused by means of an $f = 1500$ mm mirror. (**b**) The fluence profile from (**a**) along the x axis, when $y = 0$ (green curve), compared with the fluence profile of the laser field without SPC (black curve).

3.2.2. Angular Dispersion

The angular dispersion, or angular chirp (AGC), is the variation of the angle at which a specific laser field propagates with its frequency, $AGC = \frac{\partial \theta}{\partial \nu}$. For example, it can be generated when different frequency components in the spectrum are diffracted at different angles by a grating and, therefore, each of the spectral components will be tilted with different angles.

The implementation of the AGC in the code used here was performed by tilting each Gausslet with a specific propagation angle θ_m (around the axis Oy). The linear correspondence with each m component in the temporal/spectral decomposition was as in [30]: $\theta_m = AGC \cdot (\nu_0 - \nu_m)$ and the AGC value was $AGC = 1.28$ mrad/PHz on the x axis, generated, for example, by an approximately 125 µrad grating misalignment in a double-grating compressor.

In Figure 8a,b one can observe the NF pattern in the $y = 0$ and in the $x = 0$ planes, with the specific central singularity of the helical phase. The pulse front tilt associated with the AGC is present, as expected, in the $x - z$ profile. The focused field is depicted in Figure 8c,d in the focal region. The expected spatial chirp is present and can be observed as a variable front tilt in the $y = 0$ plane, in Figure 8c, along with a diagonal phase dislocation in the longitudinal profile on $y - z$ from Figure 8d.

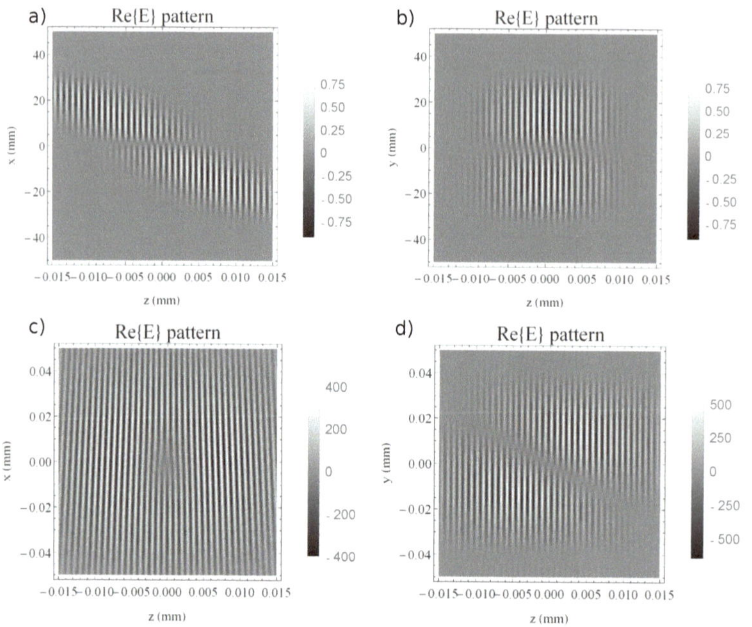

Figure 8. The electric field profile ($Re\{E\}$) of the laser field with $AGC = 1.28$ mrad/PHz in the NF region, after being reflected by the helical mirror (**a**) on $x - z$ when $y = 0$ and (**b**) on $y - z$ when $x = 0$. The $Re\{E\}$ profile for the helical field with $AGC = 1.28$ mrad/PHz in the focal region, after being focused by a mirror of $f = 1500$ mm, (**c**) on $x - z$ when $y = 0$ and (**d**) on $y - z$ when $x = 0$.

Changing the perspective, the time-integrated $|E|^2$ profile is depicted in Figure 9a. The doughnut tends to form two symmetric lobes with respect to the $y = 0$ axis. Furthermore, the field in the central hole does not drop to zero, as shown in the time-integrated profile at $y = 0$, Figure 9b. This indicates, similarly to the SPC case, a signature of the high sensitivity of the intensity profile to the presence of AGC in the helical phase pulses.

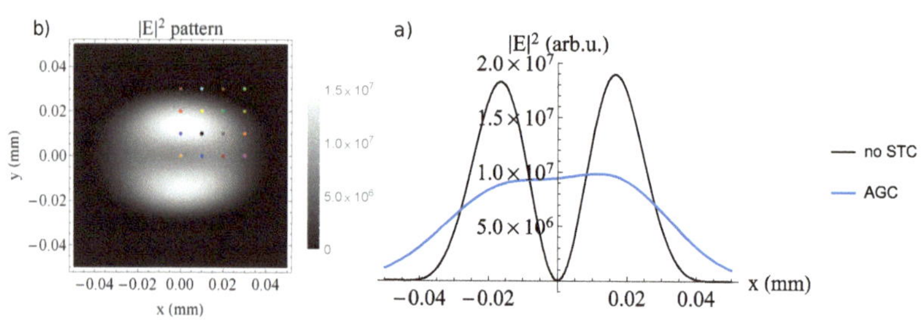

Figure 9. The fluence profile(time-integrated $|E|^2$) of the laser field with $AGC = 1.28$ mrad/PHz in the focal plane, after being reflected by the helical mirror and the focusing mirror of $f = 1500$ mm (**a**) on $x - y$ and (**b**) on x when $y = 0$ (blue curve). The AGC case is compared with the same laser field, but without STC (black curve). The colored dots in (**a**) represent the positions at which the time profiles from Figure 10 are calculated.

Furthermore, the temporal behavior of the pulse was analyzed in the focus, in Figure 10. The temporal evolution of the irradiance is depicted in the non-normalized (inset a) and normalized cases (inset b). One striking difference with respect to the ideal helical case depicted in Figure 3 is the presence of two temporal lobes. They can be symmetric or unbalanced in the field envelope. Further, the position of the maxima of the lobes shifts in time, whereas the lobe width varies, indicating the variable local pulse duration.

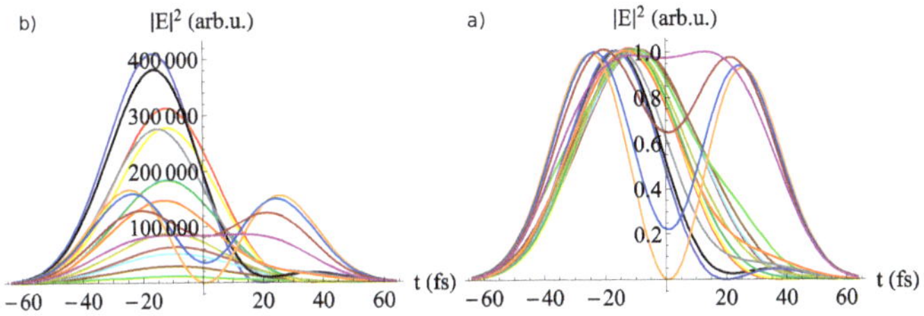

Figure 10. Thetemporal profiles of the laser field ($|E|^2$) with $AGC = 1.28$ mrad/PHz in the focal plane, after being reflected by the helical mirror and the focusing mirror of $f = 1500$ mm, (**a**) non-normalized and (**b**) normalized. The colors of the curves are the same as the ones of the dots depicted in the fluence profile in Figure 9a to indicate the positions at which they were calculated.

Several temporal envelopes can be observed in Figure 10, corresponding to horizontal cuts in the field representation of focused pulses with AGC from Figure 8c,d. This occurred due to the fact that there was a linear mapping between the propagation axis z and the temporal coordinate t. Hence, the presence of the two temporal lobes indicates the existence of the phase jump in the $y = 0$ plane, as shown in Figure 8c, whereas the shifts of the lobes and the asymmetry correspond to the tilted phase dislocation channel depicted in Figure 8d in the $x = 0$ plane.

3.2.3. Pulse Front Tilt

In lasers with STC it can happen that the wavefront and the pulse front do not coincide. When there is a linear coupling between the temporal and spatial coordinates in the formula of the laser field, this factor is known as pulse front tilt (PFT): $PFT = \frac{\partial t}{\partial x}$ [36].

In the current work, PFT was implemented as in [30] such that the central time coordinate t_{0m} (the average of each temporal Gaussian m) is shifted proportionally to the spatial position $X_{0i,j}$: $t_{0m} = PFT \cdot X_{0i,j}$. A value of $PFT = 0.8$ fs/mm was used for consistent comparison with the plots from [30].

Figure 11a,b depict the detailed field distribution in the NF in the $y = 0$ and $x = 0$ planes. These results look similar to the ones obtained in the case of AGC, presented in Figure 8. This is due to the fact that the PFT is equivalent with the AGC (in the absence of frequency chirp) [36,37]). The same qualitative behavior is also observed after the propagation of the helical PFT pulses to the focal plane, as shown in Figure 11c,d in the $y = 0$ and $x = 0$ planes.

In Figure 12, the time integrated $|E|^2$ profile is depicted for the helical pulse with $PFT = 0.8$ fs/mm, after propagation to the focal plane. Furthermore, a similar qualitative behavior is also observed here with respect to the AGC case. The depth of central hole in the beam profile is reduced in the presence of the PFT, as shown in Figure 12b, in comparison with the ideal case where no STC is present in the helical pulse.

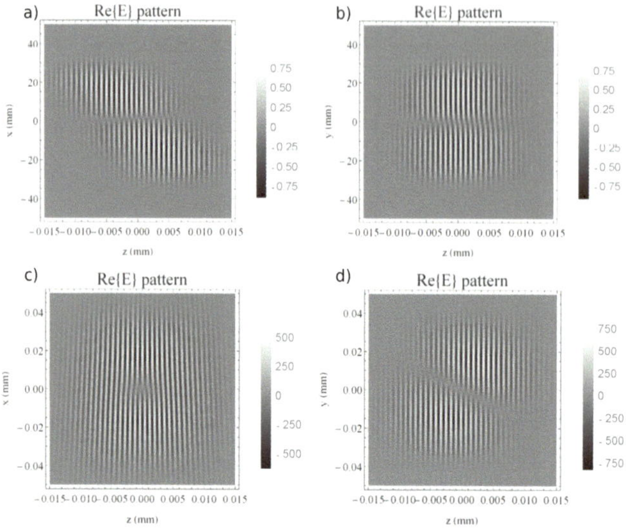

Figure 11. The electric field profile ($Re\{E\}$ at $t = 0$) of the laser field with $PFT = 0.8$ fs/mm, after being reflected by the helical mirror: in the NF (**a**) on $x - z$ for $y = 0$ and (**b**) on $y - z$ for $x = 0$; in the focus of the $f = 1500$ mm mirror (**c**) on $x - z$ for $y = 0$ and (**d**) on $y - z$ for $x = 0$. For simplicity, $z = 0$ in the focus.

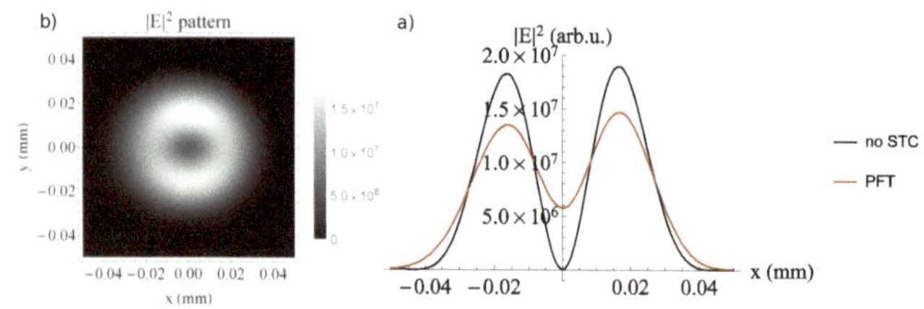

Figure 12. (**a**) The fluence profile (time-integrated $|E|^2$) on xy of the laser field with $PFT = 0.8$ fs/mm in the focal plane, after being reflected by the helical mirror and the focusing mirror of $f = 1500$ mm (**a**) in the xy plane and (**b**) along the x axis when $y = 0$ (brown curve)).

4. Discussion

In this study, the four-dimensional propagation of ultrashort optical vortices was simulated, for the first time to our knowledge, using a Gaussian decomposition code. The expected behavior in the case of non-distorted ultrashort optical vortices was obtained. Stretching and compressing the temporal shape proved that the phase displacements were maintained so that the beam profile remained of the doughnut type both in the NF and focus regions.

However, when residual STC was present in the laser field of OV, the behavior of the singularity was modified. One effect that could be easily observed in the experiments was that the central deep areain the beam profile started to wash out. As shown in Figure 7 for the case of SPC, in Figure 9 for AGC and in Figure 12 for PFT initial distortions, the amplitude of the signal in the center became significant and this can be clearly measured with a video camera. Small values of these STC distortions were enough to provide this effect, indicating a high sensitivity of the central deep amplitude of the doughnut shape.

The presence of the singularity was clearly observed in the detailed cuts in the focus areas of the SPC-, AGC- and PFT-distorted OV pulses. In order to illustrate this effect specifically for the PFT case, Figure 13 presents, at scale, a three-dimensional region plot of the pulse corresponding to the two-dimensional cuts from Figure 11c,d. It indicates that the singularity was rotated but it did not disappear. The wash out effect in the beam profile was the consequence of this rotation of the singularity orientation in the pulse along the propagation. The same rotation was responsible for the wash-out of the hole in the beam profile in the case of SPC and AGC. This rotation, even in the presence of small STC, provides additional challenges in the implementation of experiments.

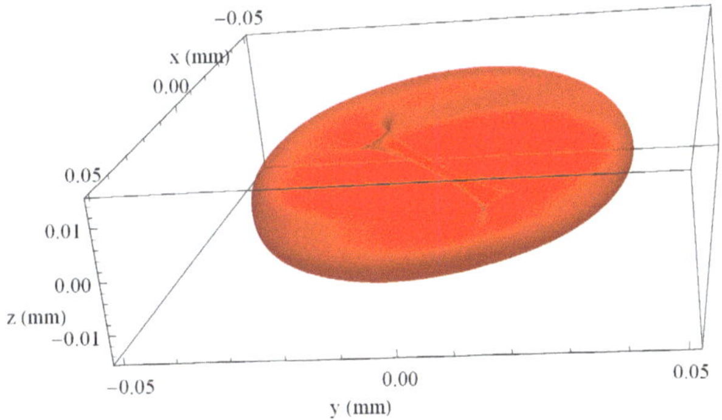

Figure 13. A 3-dimensional plot irradiance profile of the laser field ($|E|^2$) with initial $PFT = 0.8$ fs/mm, after being reflected by the helical mirror and the focusing mirror of $f = 1500$ mm, in the focus. The surface was chosen at 2% of the maximum.

Moreover, as discussed in [30], the SPC in NF generates PFT in the focus. AGC and PFT are equivalent in the absence of temporal chirp, and both of them in NF generate SPC in the focus. This behavior can also be distinguished for optical vortexes, as the field distribution is influenced by both helicity and STC. The effect of generating a specific vortex pattern in time as well, aside from the spatial effect, is an outcome of these processes [39].

Consequently, the high sensitivity of the helical pulses to the residual STCs might provide a simple path towards practical implementation of spatio-temporal optical vortices [39,40] in CPA laser systems by simply misaligning the optical stretcher or the compressor. This comes with the more subtle effects of local double pulses in the temporal profile, as shown in Figure 10.

Experiments that use helical beams—such as proton acceleration [23,41] indicating the enhanced behavior of the accelerated particles when using doughnut beams—need to take into account the effects of the STCs in the implementation and metrology phase, in order to produce the expected results. There have been many proposals of experiments using ultrashort helical pulses, e.g., in the production of gamma rays [42] and positron production [43], attosecond electron bunches with OAM [44] and relativistic electron mirrors [45], and in all these, the laser intensity distribution on the target must be optimized through systematic measurements of the STC of the vortex-free laser field.

5. Conclusions

The introduction of ultrashort laser pulses and of the CPA technique during the last quarter of the 20th century raised the need for an in-depth understanding of STC in the propagation of laser fields. This was accomplished through the development of complex metrology techniques, and also through the software development of four-dimensional propagation codes for broadband ultrashort laser pulses, such as the one used here.

In a complimentary fashion, pulse shaping techniques that implement deformable mirrors, spatial light modulators or specific optical components have enabled advances in spatially-tailored laser pulses. In particular, OVs were proposed to be used in conjunction with ultra-intense pulses from CPA laser systems in order to enhance the desired light–matter interaction effects in processes such as electron and proton acceleration. Although the impact of wavefront distortions on OVs has been reported in [31], one type of spatio-temporal distortions had not been investigated to date, to our knowledge.

A Gaussian decomposition code was used to investigate the joint presence of the OV and STC in ultrashort laser pulses and their effects in the focal plane, as required in several proposed experiments. We took as a reference the HPLS laser parameters available at ELI-NP, as these are also common to several petawatt class facilities: a Supergaussian spatial profile and a 25 fs pulse duration at a 800 nm central wavelength.

The simulations showed the evolution of the OV phase dislocation in space and time. We have also pointed out that the sensitivity of the far field to the residual STCs can help in the design of new metrology devices that provide quantitative evaluations of the STC. The results of this study will help researchers to understand better the effects of residual STCs, enabling the advanced design and implementation of future extreme light experiments with complex OV pulses.

Author Contributions: Conceptualization, D.U. and A.-M.T.; methodology, A.-M.T. and D.U.; software, A.-M.T.; validation, A.-M.T., V.I. and D.U.; formal analysis, A.-M.T.; investigation, A.-M.T. and V.I.; writing—original draft preparation, A.-M.T., V.I. and D.U.; writing—review and editing, A.-M.T., V.I. and D.U.; visualization, A.-M.T.; supervision, D.U.; project administration, D.U.; funding acquisition, D.U. All authors have read and agreed to the published version of the manuscript.

Funding: This research was funded by the Institute of Atomic Physics, Romania, grant number ELI-RO 16/2020 SBUF and by the Ministry of Education and Research, CNCS-UEFISCDI (project no. PN-IIIP4-ID-PCCF-2016–0164), within PNCDI III. The authors are thankful for the financial support from the Nuclei Project (PN 19060105). The APC was funded by PN-IIIP4-ID-PCCF-2016–0164.

Institutional Review Board Statement: Not applicable.

Informed Consent Statement: Not applicable.

Data Availability Statement: Data available on request.

Acknowledgments: The authors would like to express their appreciation to Takahisa Jitsuno, Ioan Dancus, Dan Stutman, Mihail Cernaianu, Petru Ghenuche and Dan Gh. Matei from the Extreme Light Infrastructure Nuclear Physics (ELI-NP) facility of the National Institute of Physics and Nuclear Engineering Horia Hulubei (IFIN-HH) for supporting discussions and advice. This paper is based on a presentation prepared for the '1st International Conference–Advances in 3OM: Opto-Mechatronics, Opto-Mechanics, and Optical Metrology', 13–16 December 2021, Timisoara, Romania.

Conflicts of Interest: The authors declare no conflict of interest.

Abbreviations

The following abbreviations are used in this manuscript:

AGC	Angular chirp
BPM	Beam propagation method
CPA	Chirped pulse amplification
ELI-NP	Extreme Light Infrastructure–Nuclear Physics
FDTD	Finite-difference time domain
FF	Far field
FL	Fourier limit
HPLS	High-power laser system
LG_p^l	Laguerre-Gaussian
NF	Near field
OAM	Orbital angular momentum
OPCPA	Optical parametric chirped pulse amplification
OV	Optical vortex
PFT	Pulse front tilt
PW	Petawatt
SPC	Spatial chirp
STC	Spatio-temporal couplings

References

1. Strickland, D.; Mourou, G. Compression of amplified chirped optical pulses. *Opt. Commun.* **1985**, *56*, 219–221. [CrossRef]
2. Dubietis, A.; Jonušauskas, G.; Piskarskas, A. Powerful femtosecond pulse generation by chirped and stretched pulse parametric amplification in BBO crystal. *Opt. Commun.* **1992**, *88*, 437–440. [CrossRef]
3. Lureau, F.; Matras, G.; Chalus, O.; Derycke, C.; Morbieu, T.; Radier, C.; Casagrande, O.; Laux, S.; Ricaud, S.; Rey, G.; et al. High-energy hybrid femtosecond laser system demonstrating 2 × 10 PW capability. *High Power Laser Sci. Eng.* **2020**, *8*, E43. [CrossRef]
4. Nye, J.F.; Berry, M.V. Dislocations in wave trains. *Proc. R. Soc. London A Math. Phys. Sci.* **1974**, *336*, 165–190. [CrossRef]
5. Coullet, P.; Gil, L.; Rocca, F. Optical vortices. *Opt. Commun.* **1989**, *73*, 403–408. [CrossRef]
6. Bazhenov, V.; Vasnetsov, M.; Soskin, M. Laser-Beams with Screw Dislocations in Their Wave-Fronts. *JETP Lett.* **1990**, *52*, 429–431.
7. Beijersbergen, M.; Allen, L.; van der Veen, H.; Woerdman, J. Astigmatic laser mode converters and transfer of orbital angular momentum. *Opt. Commun.* **1993**, *96*, 123–132. [CrossRef]
8. Allen, L.; Beijersbergen, M.W.; Spreeuw, R.J.C.; Woerdman, J.P. Orbital angular momentum of light and the transformation of Laguerre-Gaussian laser modes. *Phys. Rev. A* **1992**, *45*, 8185–8189. [CrossRef]
9. Grier, D.G. A revolution in optical manipulation. *Nature* **2003**, *424*, 810–816. [CrossRef]
10. Mair, A.; Vaziri, A.; Weihs, G.; Zeilinger, A. Entanglement of the orbital angular momentum states of photons. *Nature* **2001**, *412*, 313–316. [CrossRef]
11. Molina-Terriza, G.; Torres, J.P.; Torner, L. Twisted photons. *Nat. Phys.* **2007**, *3*, 305–310. [CrossRef]
12. Devlin, R.C.; Ambrosio, A.; Rubin, N.A.; Mueller, J.P.B.; Capasso, F. Arbitrary spin-to–orbital angular momentum conversion of light. *Science* **2017**, *358*, 896–901. [CrossRef] [PubMed]
13. Stav, T.; Faerman, A.; Maguid, E.; Oren, D.; Kleiner, V.; Hasman, E.; Segev, M. Quantum entanglement of the spin and orbital angular momentum of photons using metamaterials. *Science* **2018**, *361*, 1101–1104. [CrossRef] [PubMed]
14. Gong, M. Recent Advances on Tunable Vortex Beam Devices for Biomedical Applications. *Biomed. J. Sci. Tech. Res.* **2018**, *9*, 5. [CrossRef]
15. Fuerhapter, S.; Jesacher, A.; Bernet, S.; Ritsch-Marte, M. Spiral phase contrast imaging in microscopy. *Opt. Express* **2005**, *13*, 689–694. [CrossRef]
16. Tamburini, F.; Anzolin, G.; Umbriaco, G.; Bianchini, A.; Barbieri, C. Overcoming the Rayleigh Criterion Limit with Optical Vortices. *Phys. Rev. Lett.* **2006**, *97*, 163903. [CrossRef]
17. Hell, S.W. Microscopy and its focal switch. *Nat. Methods* **2008**, *6*, 24–32. [CrossRef]
18. Barreiro, J.T.; Wei, T.C.; Kwiat, P.G. Beating the channel capacity limit for linear photonic superdense coding. *Nat. Phys.* **2008**, *4*, 282–286. [CrossRef]
19. Lee, J.C.T.; Alexander, S.J.; Kevan, S.D.; Roy, S.; McMorran, B.J. Laguerre–Gauss and Hermite–Gauss soft X-ray states generated using diffractive optics. *Nat. Photonics* **2019**, *13*, 205–209. [CrossRef]
20. De las Heras, A.; Pandey, A.K.; Román, J.S.; Serrano, J.; Baynard, E.; Dovillaire, G.; Pittman, M.; Durfee, C.G.; Plaja, L.; Kazamias, S.; et al. Extreme-ultraviolet vector-vortex beams from high harmonic generation. *Optica* **2022**, *9*, 71–79. [CrossRef]
21. Allen, L.; Padgett, M.; Babiker, M. The Orbital Angular Momentum of Light. In *Progress in Optics*; Elsevier: Amsterdam, The Netherlands, 1999; pp. 291–372. [CrossRef]
22. Nuter, R.; Korneev, P.; Dmitriev, E.; Thiele, I.; Tikhonchuk, V.T. Gain of electron orbital angular momentum in a direct laser acceleration process. *Phys. Rev. E* **2020**, *101*, 053202. [CrossRef] [PubMed]
23. Brabetz, C.; Busold, S.; Cowan, T.; Deppert, O.; Jahn, D.; Kester, O.; Roth, M.; Schumacher, D.; Bagnoud, V. Laser-driven ion acceleration with hollow laser beams. *Phys. Plasmas* **2015**, *22*, 013105. [CrossRef]
24. Dennis, M.R.; O'Holleran, K.; Padgett, M.J. Chapter 5 Singular Optics: Optical Vortices and Polarization Singularities. In *Progress in Optics*; Elsevier: Amsterdam, The Netherlands, 2009; pp. 293–363. [CrossRef]
25. Pariente, G.; Gallet, V.; Borot, A.; Gobert, O.; Quéré, F. Space–time characterization of ultra-intense femtosecond laser beams. *Nat. Photonics* **2016**, *10*, 547–553. [CrossRef]
26. Dorrer, C. Spatiotemporal Metrology of Broadband Optical Pulses. *IEEE J. Sel. Top. Quantum Electron.* **2019**, *25*, 1–16. [CrossRef]
27. Greynolds, A.W. Propagation Of Generally Astigmatic Gaussian Beams Along Skew Ray Paths. In *Proceedings of the Diffraction Phenomena in Optical Engineering Applications*; Byrne, D.M., Harvey, J.E., Eds.; International Society for Optics and Photonics, SPIE: Bellingham, WA, USA, 1986; Volume 0560, pp. 33–51. [CrossRef]
28. Harvey, J.E.; Irvin, R.G.; Pfisterer, R.N. Modeling physical optics phenomena by complex ray tracing. *Opt. Eng.* **2015**, *54*, 035105. [CrossRef]
29. Worku, N.G.; Gross, H. Gaussian pulsed beam decomposition for propagation of ultrashort pulses through optical systems. *J. Opt. Soc. Am. A* **2020**, *37*, 98–107. [CrossRef]
30. Talposi, A.M.; Ursescu, D. Propagation of ultrashort laser fields with spatiotemporal couplings using Gabor's Gaussian complex decomposition. *J. Opt. Soc. Am. A* **2022**, *39*, 267–278. [CrossRef]
31. Ohland, J.B.; Eisenbarth, U.; Roth, M.; Bagnoud, V. A study on the effects and visibility of low-order aberrations on laser beams with orbital angular momentum. *Appl. Phys. B* **2019**, *125*, 202. [CrossRef]
32. Wolfram Research. NonlinearModelFit, Wolfram Language Function. 2008. Available online: https://reference.wolfram.com (accessed on 30 September 2021).

33. Siegman, A. *Lasers*; University Science Books: Sausalito, CA, USA, 1986; Chapter 17.
34. Serna, J.; Encinas-Sanz, F.; Nemes, G. Complete spatial characterization of a pulsed doughnut-type beam by use of spherical optics and a cylindrical lens. *J. Opt. Soc. Am. A* **2001**, *18*, 1726–1733. [CrossRef]
35. Zhou, G.; Cai, Y.; Dai, C.Q. Hollow vortex Gaussian beams. *Sci. China Phys. Mech. Astron.* **2013**, *56*, 896–903. [CrossRef]
36. Akturk, S.; Gu, X.; Gabolde, P.; Trebino, R. The general theory of first-order spatio-temporal distortions of Gaussian pulses and beams. *Opt. Express* **2005**, *13*, 8642–8661. [CrossRef] [PubMed]
37. Akturk, S.; Gu, X.; Zeek, E.; Trebino, R. Pulse-front tilt caused by spatial and temporal chirp. *Opt. Express* **2004**, *12*, 4399–4410. [CrossRef] [PubMed]
38. Sukhorukov, A.P.; Yangirova, V.V. Spatio-temporal vortices: Properties, generation and recording. In *Proceedings of the Nonlinear Optics Applications*; Karpierz, M.A., Boardman, A.D., Stegeman, G.I., Eds.; International Society for Optics and Photonics, SPIE: Bellingham, WA, USA, 2005; Volume 5949, pp. 35–43. [CrossRef]
39. Hancock, S.W.; Zahedpour, S.; Goffin, A.; Milchberg, H.M. Free-space propagation of spatiotemporal optical vortices. *Optica* **2019**, *6*, 1547. [CrossRef]
40. Chong, A.; Wan, C.; Chen, J.; Zhan, Q. Generation of spatiotemporal optical vortices with controllable transverse orbital angular momentum. *Nat. Photonics* **2020**, *14*, 350–354. [CrossRef]
41. Wang, W.; Jiang, C.; Dong, H.; Lu, X.; Li, J.; Xu, R.; Sun, Y.; Yu, L.; Guo, Z.; Liang, X.; et al. Hollow plasma acceleration driven by a relativistic reflected hollow laser. *Phys. Rev. Lett.* **2020**, *125*, 034801. [CrossRef]
42. Zhang, H.; Zhao, J.; Hu, Y.; Li, Q.; Lu, Y.; Cao, Y.; Zou, D.; Sheng, Z.; Pegoraro, F.; McKenna, P.; et al. Efficient bright γ-ray vortex emission from a laser-illuminated light-fan-in-channel target. *High Power Laser Sci. Eng.* **2021**, *9*, 1–24. [CrossRef]
43. Zhao, J.; Hu, Y.T.; Lu, Y.; Zhang, H.; Hu, L.X.; Zhu, X.L.; Sheng, Z.M.; Turcu, I.C.E.; Pukhov, A.; Shao, F.Q.; et al. All-optical quasi-monoenergetic GeV positron bunch generation by twisted laser fields. *Commun. Phys.* **2022**, *5*, 15. [CrossRef]
44. Hu, L.X.; Yu, T.P.; Sheng, Z.M.; Vieira, J.; Zou, D.B.; Yin, Y.; McKenna, P.; Shao, F.Q. Attosecond electron bunches from a nanofiber driven by Laguerre-Gaussian laser pulses. *Sci. Rep.* **2018**, *8*, 7282. [CrossRef]
45. Hu, L.X.; Yu, T.P.; Li, H.Z.; Yin, Y.; McKenna, P.; Shao, F.Q. Dense relativistic electron mirrors from a Laguerre-Gaussian laser-irradiated micro-droplet. *Opt. Lett.* **2018**, *43*, 2615. [CrossRef]

photonics

MDPI

Article

Low-Cost 3D-Printed Electromagnetically Driven Large-Area 1-DOF Optical Scanners

Ching-Kai Shen [1], Yu-Nung Huang [1], Guan-Yang Liu [1], Wei-An Tsui [1], Yi-Wen Cheng [1], Pin-Hung Yeh [1] and Jui-che Tsai [1,2,*]

[1] Graduate Institute of Photonics and Optoelectronics, National Taiwan University, Taipei 10617, Taiwan; d09941019@ntu.edu.tw (C.-K.S.); r08941111@ntu.edu.tw (Y.-N.H.); r09941067@ntu.edu.tw (G.-Y.L.); r09941075@ntu.edu.tw (W.-A.T.); r09941083@ntu.edu.tw (Y.-W.C.); r09941158@ntu.edu.tw (P.-H.Y.)

[2] Department of Electrical Engineering, National Taiwan University, Taipei 10617, Taiwan

* Correspondence: juichetsai@ntu.edu.tw

Abstract: In this paper, we demonstrate 3D-printed 1-DOF (one torsional axis; 1 degree of freedom) optical scanners with large mirror areas (up to 20×20 mm^2). Each device consists of an aluminum-coated square silicon substrate serving as the mirror, two miniature permanent magnets, an electromagnet, and a 3D-printed structure including the mirror frame, torsion springs, and base. One device can reach a static half optical scan angle of 14.8 deg., i.e., a full optical scan angle of 29.6 deg., at 12 VDC; this particular device exhibits a mechanical resonance frequency of 84 Hz. These scanners can be a potential, low-cost alternative to the expensive conventional galvanometer scanners.

Keywords: electromagnetically driven; 1-DOF scanning mirrors; 3D printing

1. Introduction

Optical scanners have been frequently found in various types of optical systems. Among them, the galvanometer scanners [1,2] are perhaps most popular thanks to their commercial availability and reasonable pricing; outstanding efforts have also been put into analyzing and optimizing the scanning functions [2,3]. They can be found widely in the market [4–6]. Their sizes fall within the range from a few millimeters to a few centimeters. Some of them can even reach an optical scan angle of 40 degrees [4]. They are commonly seen in early optical coherence tomography (OCT) systems, either to direct the light beam to the sample [7], or to generate the necessary optical pathlength difference for depth scanning [8]. Other applications include laser scanning microscopy, LiDAR, etc. [9,10]. Although affordable, the cost ranges from a few thousand US dollars to more than ten US dollars [5]; therefore, galvanometer scanners are usually not considered as a low-cost option. Particularly, at the developing phase of a system or experiment that involves galvanometer scanners, the required specifications of the scanners may still be unclear. The price tags of the galvanometer scanners prevent researchers from moving rapidly forwards for a practical trial; contrarily, they spend much effort carefully designing "on paper" in advance just to make sure they get the right scanners later for their systems. This philosophy exists in most research groups in academia and industry, and somewhat slow down the technology development progress.

Riding on the shoulders of the MEMS fabrication technologies, various MEMS optical scanners have been developed. The common driving mechanisms of MEMS scanners include electrostatic actuation [11,12], electromagnetic actuation [13,14], electrothermal actuation [15,16], and piezoelectric actuation [17]. Many electrostatic micromirrors require undesirable high voltages that sometimes approach a hundred volts [18]; electrothermal actuation and electromagnetic actuation both lead to direct heating of the mirror, which can potentially change the mirror parameters such as the curvature; piezoelectric material is less popular in semiconductor fabrication business. Despite these shortcomings, MEMS scanners (mostly micromirrors) make it feasible to accomplish system miniaturization and portability. The fabrication normally requires cleanroom usage and the extremely expensive and heavy equipment such as the etching and deposition machines. Their costs can be lowered by using wafer-level fabrication processes for mass production. However, this is only possible with mature products of high demand; in other words, academia or business developing new products mostly still suffers the high cost per device resulting from the cleanroom microfabrication process. Moreover, the typical sizes of MEMS scanners are between a few hundred microns and a few millimeters, which makes them unsuitable for handling light beams with large cross sections; this limits the number of resolvable spots [19] in a laser scanning system such as the laser scanning microscope and laser scanning projector.

Prism scanners belong to another category which has drawn a great deal of attention [20–22]. Distinct from mirrors, prims are transparent optical components whereby the light paths do not need to be folded, making it easier to design and construct the optical systems. Beautiful scan patterns were previously presented using rotational Risley prisms [20]. The only concern about prism scanners is the prisms' bulky sizes, which make fast scanning somewhat challenging.

In this paper, we demonstrate 3D-printed optical scanners with large mirror areas (up to 20×20 mm^2). The main mechanical structure of the device is fabricated by FDM (fused deposition modeling) 3D printing; an aluminum-coated silicon chip is mounted on the top to serve as the reflecting surface. Each scanner is equipped with miniature permanent magnets and driven by an electromagnet; both the permanent magnets and electromagnet are off the shelf with extremely low prices. No electric current is delivered to the mirror so there is no issue of heating the mirror. The devices presented in this paper exhibit low driving voltages and are positioned to be an extremely low-cost alternative to the expensive conventional galvanometer scanners. The estimated cost per device is less than ten US dollars. Furthermore, thanks to the rapid prototyping of 3D printing, the turnaround time to complete fabrication and delivery is <1 week.

2. Device Design and Fabrication

We propose using FDM (fused deposition modeling) 3D printing to fabricate our devices. FDM 3D printing possesses the advantages of low cost, high speed, and large area printing. Compared to other printing techniques such as stereolithography (SLA) 3D printing and selective laser sintering (SLS) 3D printing, it is more environment-friendly because it avoids the use of the photocurable resin and polymer powder. The printed material is polylactic acid (PLA), which is low-cost and environment-friendly; the printer used in this study is Ultimaker 2+.

Each device consists of an aluminum-coated square silicon substrate serving as the mirror, two miniature permanent magnets, an electromagnet, and a 3D-printed structure including the mirror frame, torsion springs, and base; if needed, a soft iron core can be used to extend the magnetic field produced by the electromagnet. As shown in Figure 1, the exploded view drawing of the device, the two miniature permanent magnets are attached beneath the mirror frame, oriented in opposite directions. When an electric current is sent through the electromagnet, attracting and repelling forces and, therefore, a torque are generated to rotate the mirror. The magnetic flux density generated by the electromagnet is 200 G at 5 V.

The two practical arrangements are shown in Figure 2, one without the soft iron core for extension (type I) and the other with it (type II). For type I devices, the electromagnet generates a magnetic field to directly attract or repel the miniature permanent magnets (oriented in opposite directions) under the mirror frame to drive the mirror to rotate. For type II devices, a soft iron core is connected to the electromagnet to make the magnetic field extend upward along the soft iron core. In this way, the distance between the electromagnet and the miniature permanent magnets can be increased.

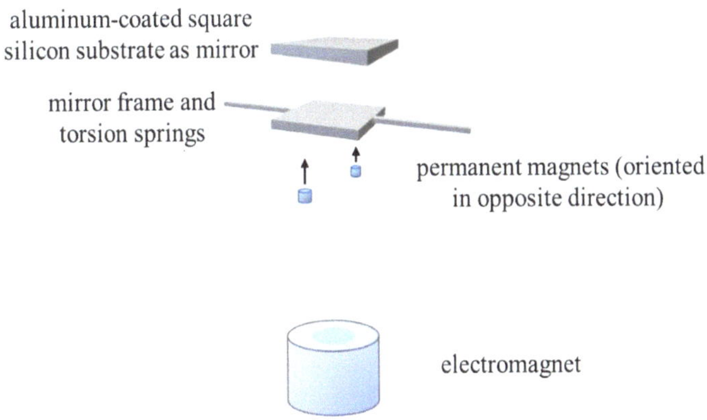

Figure 1. Exploded view drawing of the device.

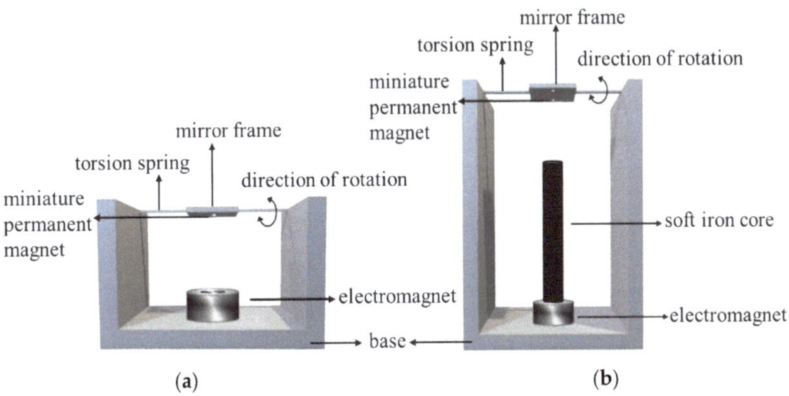

Figure 2. Two practical arrangements: (**a**) Type I without the soft iron core for extension; (**b**) Type II with the soft iron core for extension.

The main structure of the device is made by FDM 3D printing. An aluminum-coated silicon chip (slice) is mounted on the frame to serve as the mirror. Then, the miniature permanent magnets are attached to the back of the mirror frame, and the electromagnet is placed on the base.

In our study, five different designs are employed. Their schematic 3D drawings (not to scale) and designed dimensions can be found in Table 1. Table 2 summarizes the specifications of all of the designs. Figure 3 shows the photos of the actual fabricated devices.

Table 1. Schematic 3D drawings and designed dimensions of the five different designed devices.

	Oblique Top View	Side View	Bottom View (Beneath the Mirror)
A (Type I)	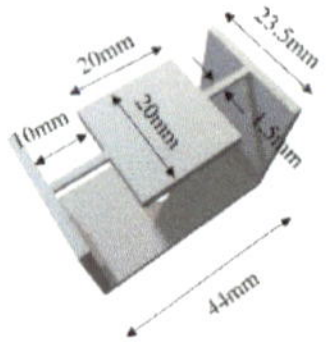	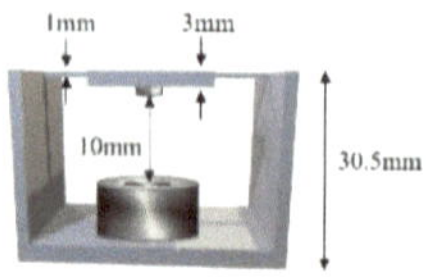	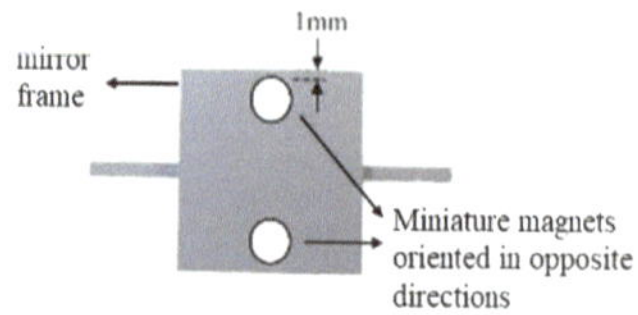
B (Type I)	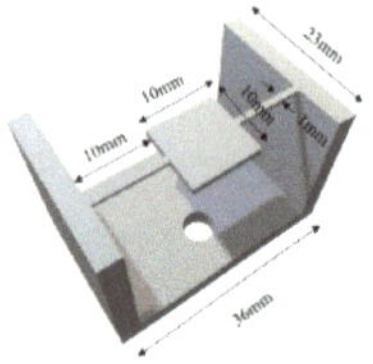	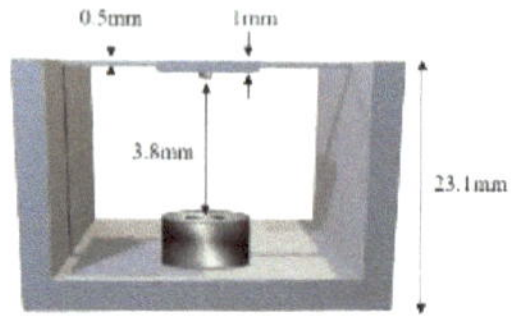	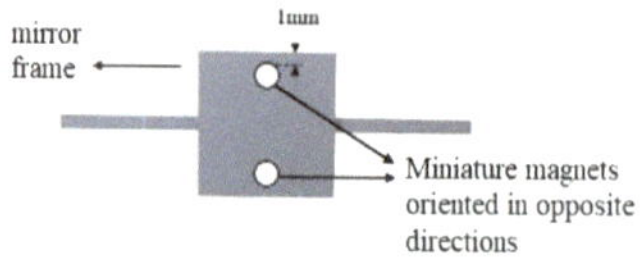
C (Type II)	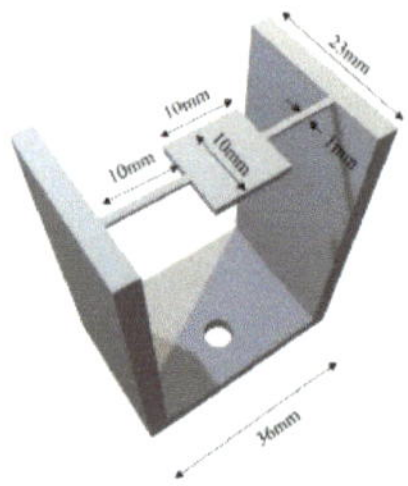	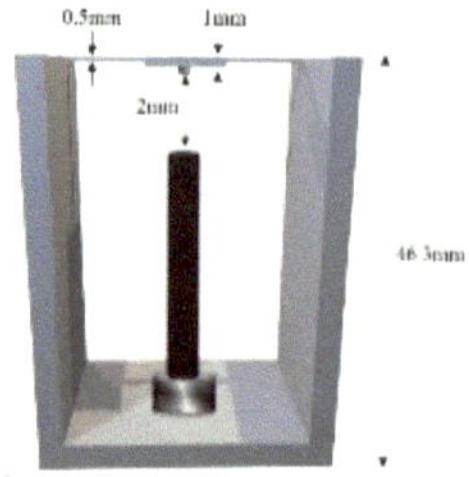	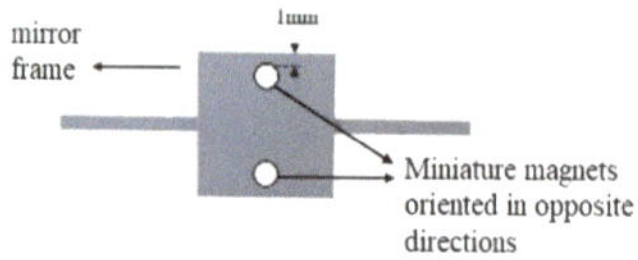
D (Type I)	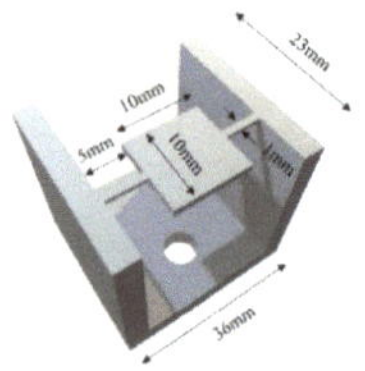	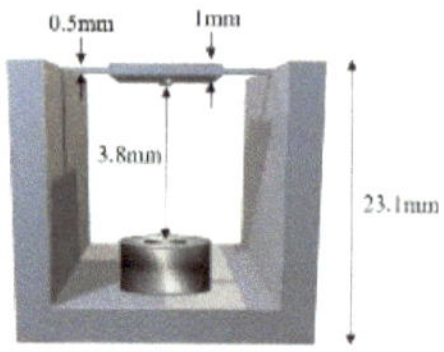	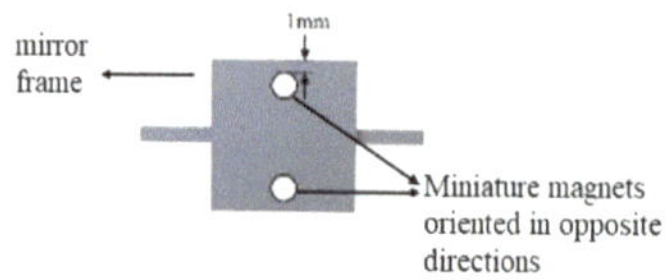
E (Type II)	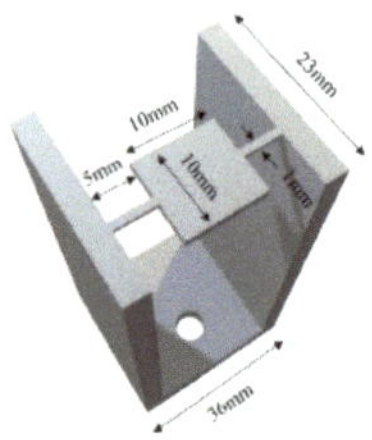	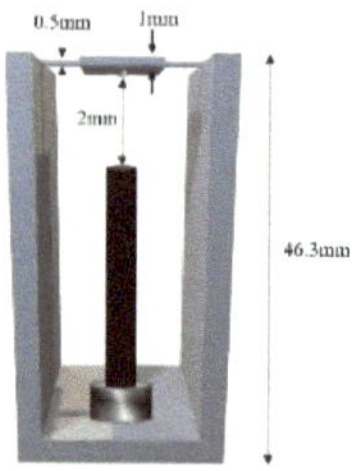	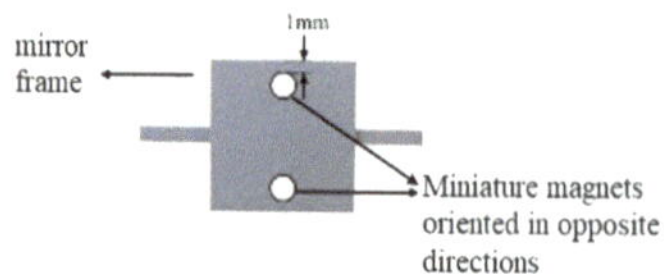

Table 2. Specifications of the designed devices.

	A (Type I)	B (Type I)	C (Type II)	D (Type I)	E (Type II)
Mirror frame size (mm^3)	20 × 20 × 3	10 × 10 × 1	10 × 10 × 1	10 × 10 × 1	10 × 10 × 1
Torsion spring size (mm^3)	10 × 1.5 × 1	10 × 1 × 0.5	10 × 1 × 0.5	5 × 1 × 0.5	5 × 1 × 0.5

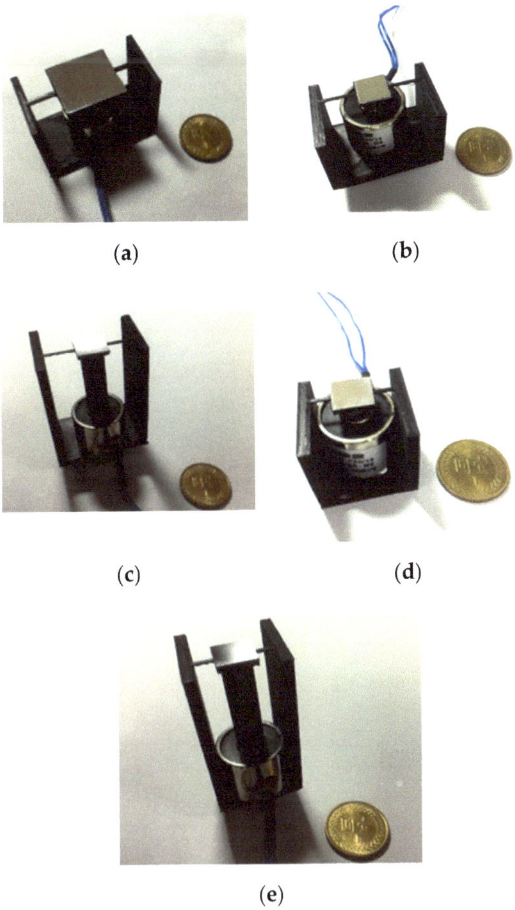

Figure 3. The photos of the fabricated devices: (**a**) Device A (**b**) Device B (**c**) Device C (**d**) Device D (**e**) Device E.

3. Device Characterization and Results

3.1. Experiment Setup and Characterization

The device characterization mainly focuses on measuring the scan angles of all of the designs. Specifically, for each design we will experimentally find out the static optical deflection angles at different applied dc voltages, i.e., the dc characteristic. The scanner will also be driven with a sinusoidal voltage for dynamic characterization; the frequency of the sinusoidal waveform will be tuned and the scan range at each frequency will be recorded; this measurement can lead to the discovery of the frequency response and resonance frequency.

Figure 4a is the schematic drawing of the experimental setup used to obtain the dc characteristic. A 632.8 nm HeNe laser is incident on the mirror and reflected toward a screen. The power supply provides a dc voltage which is applied to the electromagnet while the oscilloscope measures the exact voltage drop across the electromagnet. When the voltage is tuned, the reflected laser spot on the screen moves and the displacement d is recorded. As shown in Figure 4b, given the distance L between the scanner and screen, the optical deflection angle at a certain voltage is then $\tan^{-1}(d/L)$, which is considered as the optical half angle θ_{half}. Flipping the voltage polarity overturns the polarity of the electromagnet, and rotates the mirror toward the opposite direction; therefore, in the ideal case the full scan range of the scanner is twice the maximum optical half angle, i.e., $2\theta_{half}$, max.

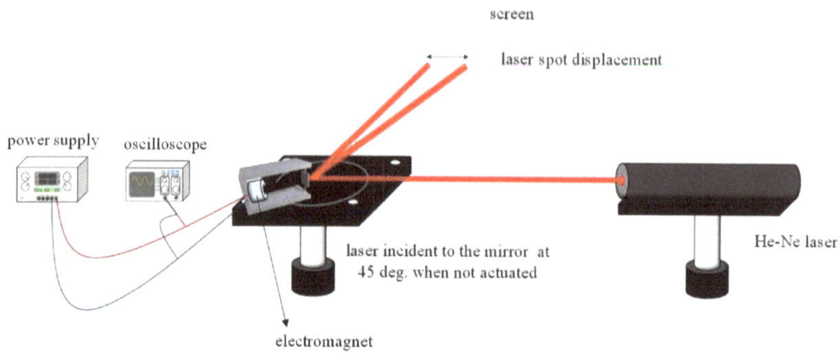

(**a**)

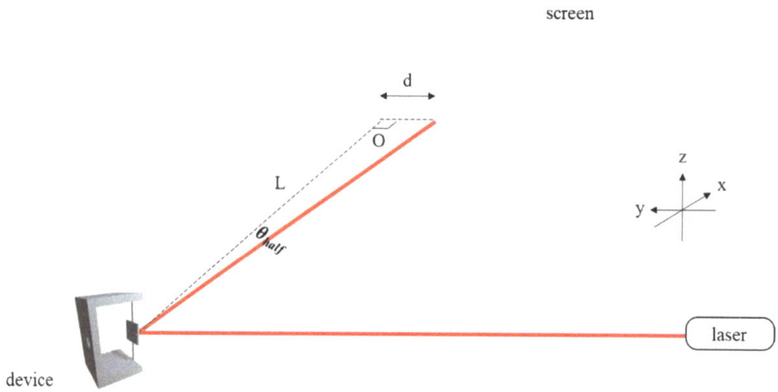

(**b**)

Figure 4. dc characterization: (**a**) experimental setup; (**b**) the optical half angle (θ_{half}) can be found using the given L and measured d.

The measurement setup for dynamic characterization is similar to that for dc characterization, as shown in Figure 5a, with the replacement of the power supply by a function generator. With a sinusoidal voltage applied to the electromagnet, the electromagnet's polarity flips periodically; this enables the mirror to swing to the two extremes alternately. As shown in Figure 5b, $\theta_{full} = 2\ \theta_{half} = 2\tan^{-1}(d/2L)$ is the full optical scan range at a certain voltage amplitude and frequency. To find the frequency response of the device, the frequency of the sinusoidal waveform is swept manually while keeping the amplitude across the electromagnet constant, and the full optical scan ranges at several frequencies are recorded.

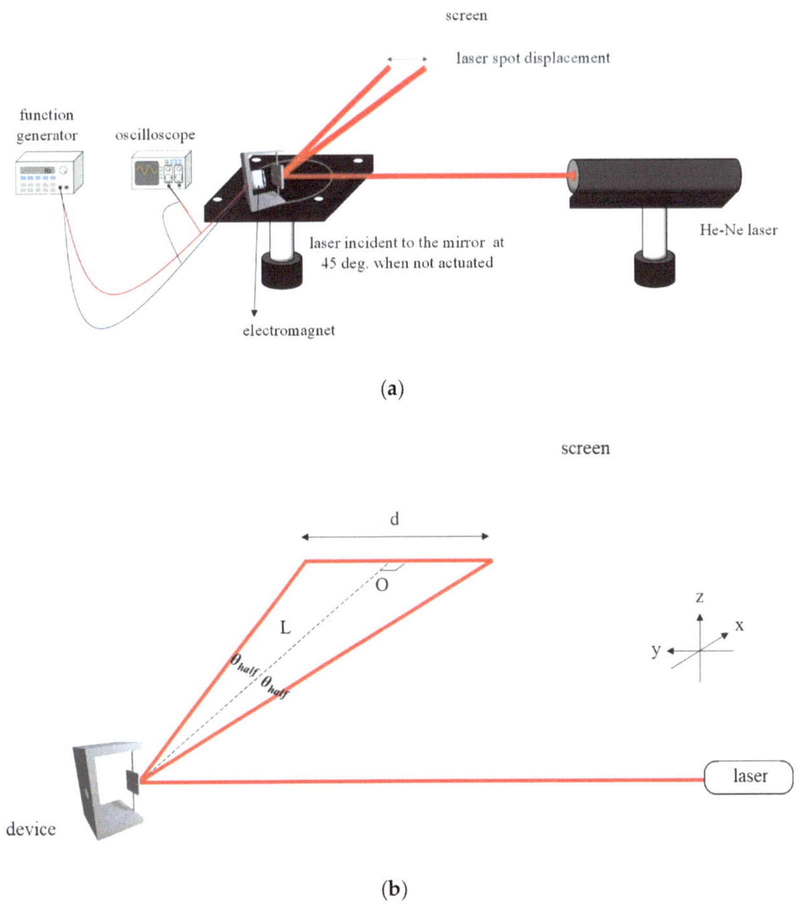

Figure 5. Dynamic characterization: (**a**) experimental setup; (**b**) the full optical scan range [θ_{full} = 2 θ_{half} = 2tan^{-1}(d/2L)] at a certain voltage amplitude and frequency can be found using the given L and measured d; the frequency response can be obtained by sweeping the frequency of the function generator.

3.2. Experiment Results

Figure 6 shows the static characteristics, optical half angle vs. voltage, of all of the devices. Among all of the devices, Device A has the largest mirror size (20 × 20 mm^2) and does not have the soft iron core for extension; it exhibits the largest optical deflection angle. At a 12 V dc voltage, the optical half angle θ_{half} is as large as 14.8 degrees, which means the full scan range of this scanner is 29.6 degrees under quasi-static operation. This is a considerable angular range for a mirror-type beam scanner.

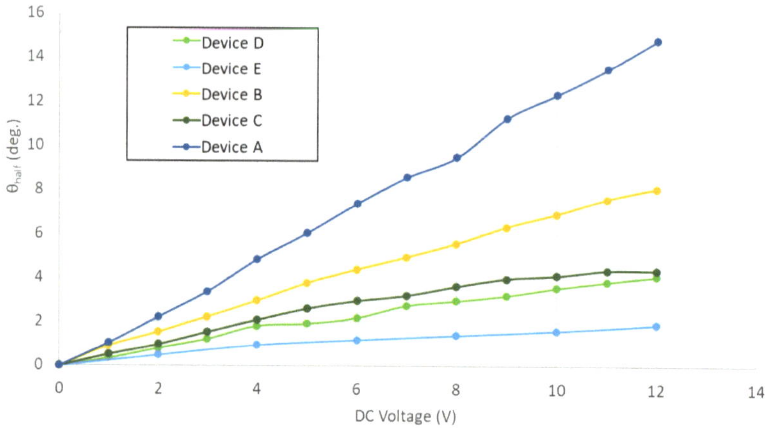

Figure 6. Static characteristics: optical half angle θ_{half} vs. dc voltage for different devices.

Figure 7 is the frequency response of Device D. The peak-to-peak voltage is kept a constant when the frequency is swept. Based on the frequency response curve, when the driving frequency is 341 Hz, the full optical scan range θ_{full} reaches the maximum; 341 Hz is then the resonance frequency of the device.

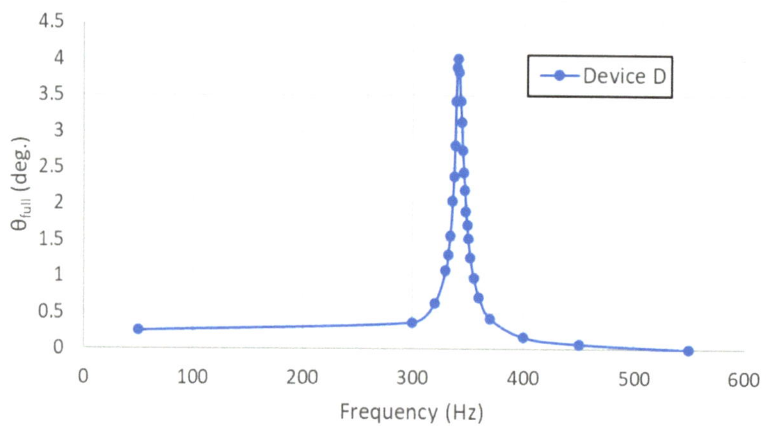

Figure 7. The frequency response of Device D.

Figure 8 shows the traces of the reflected light spot on the screen of Device D at several selected frequencies, including the resonance frequency 341 Hz. It can be seen that the longest trace does occur at the resonance frequency. No rotation around the orthogonal axis is observed.

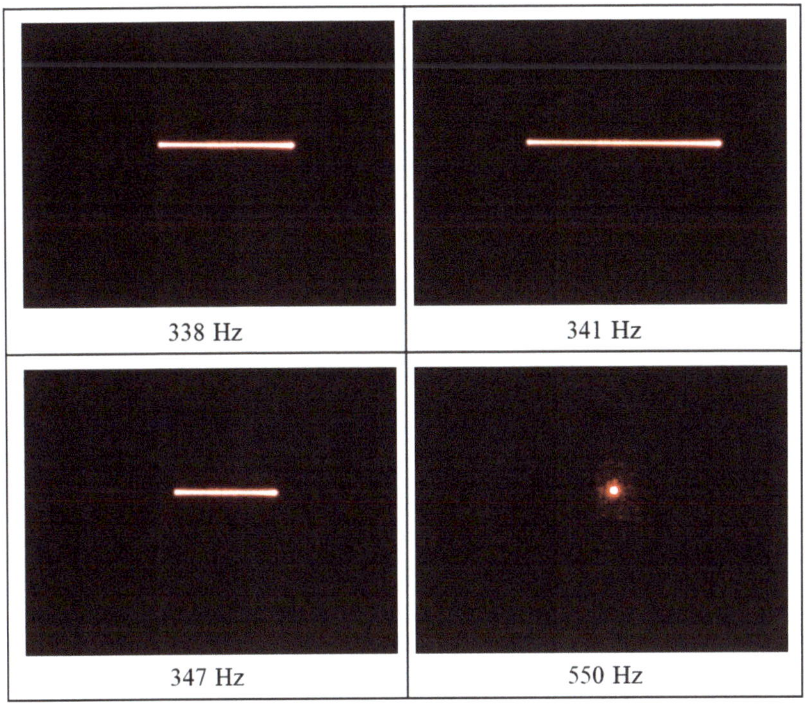

Figure 8. Traces of the reflected light spot on the screen of Device D at several selected frequencies, including the resonance frequency 341 Hz.

Table 3 lists the measured mirror frame thickness, spring size, and torsional resonance frequency of each device, as well as the estimated mirror mass (including the silicon slice, coated aluminum, and permanent magnets), estimated torsion and bending spring constants, and predicted torsional and bending resonance frequencies. When calculating the spring constants and mirror mass and predicting the resonance frequencies, we take the measured dimensions instead of the designed values.

Table 3. Estimated and measured parameters of each device.

Devices	Measured Mirror Frame Thickness (mm)	Measured Spring Size (mm^3, Length × Width × Thickness)	Est. Mirror Mass (g)	Est. Torsion Spring Const. (N m/rad)	Est. Bending Spring Const. (N/m)	Predicted Torsional Resonance Freq. (Hz)	Predicted Bending Resonance Freq. (Hz)	Measured Torsional Resonance Freq. (Hz)
Device A	2.57	9.82 × 1.6 × 0.99	2.58	3.24×10^{-2}	4460	91.25	209.19	84
Device B	1.13	9.97 × 1.01 × 0.65	0.279	5.56×10^{-3}	761.2	243.94	263.02	229
Device C	1.12	9.83 × 1.1 × 0.62	0.278	5.76×10^{-3}	750.6	248.66	261.77	227
Device D	1.16	4.9 × 1.02 × 0.59	0.271	9.1×10^{-3}	4844	316.42	673.05	341
Device E	1.15	4.87 × 1.02 × 0.62	0.281	10.34×10^{-3}	5724	331	718.07	348

As expected, the resonance frequency is mainly related to the mirror geometry and torsion springs. Whether the electromagnet has an iron core for extension or not should not affect the resonance frequency.

The predicted torsion spring constant k_t and resonance frequency f_t are calculated using the following equations:

$$k_t = 2K \times \frac{G}{L} \quad (1)$$

$$f_t = \frac{1}{2\pi}\sqrt{\frac{k_t}{I}} \quad (2)$$

where G is the shear modulus of PLA, L is the length of the spring, I is the rotational moment of inertia of the mirror, and K is the shape factor of the spring's cross section which can be expressed as

$$\mathrm{K} \approx \mathrm{a}b^3\left[\frac{16}{3} - 3.36\frac{b}{a}\left(1 - \frac{b^4}{12a^4}\right)\right] \tag{3}$$

where a is half of the spring width, and b is half of the spring thickness [23].

The predicted bending spring constant k_b and resonance frequency f_b are calculated with:

$$k_b = \frac{32Eab^3}{L^3} \tag{4}$$

$$f_b = \frac{1}{2\pi}\sqrt{\frac{k_b}{m}} \tag{5}$$

where E is the Young's modulus of PLA, and m is the mass of the mirror. The Young's modulus and mass density of PLA chosen for our calculation are 1.36 GPa and 1240 $\mathrm{kg/m^3}$, respectively. We will present a further discussion regarding the calculation in the next section.

4. Discussion

The PLA's Young's modulus can vary within a wide range [24]. Apparently, the modulus of our device is on the lower side. We assume 1.36 GPa as the Young's modulus, and the difference between the prediction and actual performance of the torsion mode is smaller than 10%. With the Young's modulus and mass density set to 1.36 GPa and 1240 $\mathrm{kg/m^3}$, respectively, and with the measured spring dimensions and mirror frame's thickness, we are able to calculate/estimate/predict the resonance frequencies of the bending modes, torsional and bending spring constants, and mass, which are all included in Table 3 above.

In Devices A, D, and E, the bending mode exhibits a frequency far higher than that of the torsion mode, so coupling should be insignificant. In Devices B and C, the difference between the bending and torsion mode frequencies is not as large, but it is still greater than the measured resonance mode width of the torsion mode; therefore, coupling should also be negligible. Furthermore, the two permanent magnets of each mirror are oriented in opposite directions so the net force on the mirror should be minimized, if not zero, making it hard to excite the bending mode significantly. In summary, coupling should be insignificant, and it is unlikely to excite a substantial bending motion. We did verify this experimentally. In our experiments (referring to Figure 5 which shows the setup to find the resonance), we found that d increased with L as expected at any tested frequency, which would not have been possible if the out-of-plane motion, instead of rotation, had been excited; an out-of-plane motion would have resulted in a lateral shift of the laser beam instead of a deflection angle. Moreover, a significant out-of-plane motion would cause the laser beam to miss the mirror, which was never observed during the experiments.

Regarding the effect of any possible misalignment of the permanent magnets, if we take Device A as an example, misalignments of 0.5 mm of the permanent magnets may result in an estimated 7% change in the static angle. In Devices B–E, such misalignments of the permanent magnets may result in an estimated 14% change in the static angle.

Another concern is the consistency between prints, and we can infer from the measured resonance frequencies. Devices B and C have the same mirror frame design, while Devices D and E share the same mirror frame design. As shown in Table 3, the measured resonant frequencies of Devices B and C are 229 and 227 Hz, respectively, i.e., <0.9% difference; the measured resonant frequencies of Devices D and E are 341 and 348 Hz, respectively, i.e., 2% difference. This demonstrates decent consistency between prints.

In terms of device lifetime, the parameters of the printed material can indeed change with time because of environmental factors, such as humidity and temperature. One of our future plans is to perform a long-term monitoring of the devices to study the effect of those factors. From a different point of view, our low-cost devices can be disposable had they become unusable due to those long-term effects from the environmental factors; the rapid prototyping nature of 3D printing can provide fast delivery of a new device.

Our goal is to propose an extremely low-cost scanner solution. The 3D printer costed about 4000 USD, and our estimated material cost per scanner is merely <10 USD (PLA: <0.4 USD; silicon slice + aluminum coating: <2.5 USD; electromagnet: <6.5 USD; permanent magnets: <0.2 USD); no other charge was incurred.

Closed-loop control has been a popular approach for improving scanner performance [25]. However, since we position our work as an extremely low-cost scanner solution, we intentionally avoid the closed-loop control to keep the cost low. We have tested a scanner with a modified but similar design under open-loop operation using a PSD (position sensing detector); the results show good angular repeatability with an error of only ~1.6% (Figure 9).

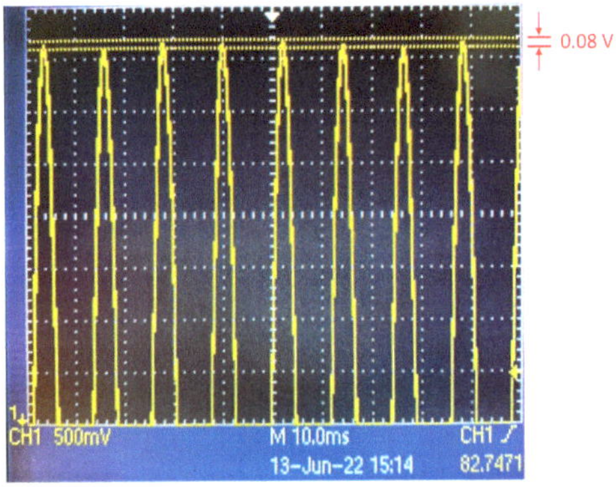

Figure 9. PSD output of the scanner angular repeatability test. The output voltage amplitude is ~5 V with variation of <0.08 V, i.e., ~1.6%.

A scanner can be dedicated to either quasi-static scanning or dynamic scanning. The static characterization demonstrated in Section 3 has proven the ability of our scanners to perform quasi-static scanning. Dynamic scanning includes raster scanning [14], spiral scanning [26], Lissajous scanning [27], etc. Our scanners exhibit static optical full angles of >4 deg.; Devices A and B can even reach 29 deg. and 16 deg., respectively. Moreover, all of the measured resonance frequencies are greater than 60 Hz. Therefore, our devices can serve as the slow-axis scanner for raster scanning with a 60 Hz frame rate. As for Lissajous scanning, using a pair of our revised scanners with minor modification (mainly to increase the mirror area), we are able to generate a nice Lissajous pattern (Figure 10). The frequencies of the x and y scans are 70 and 58 Hz, respectively.

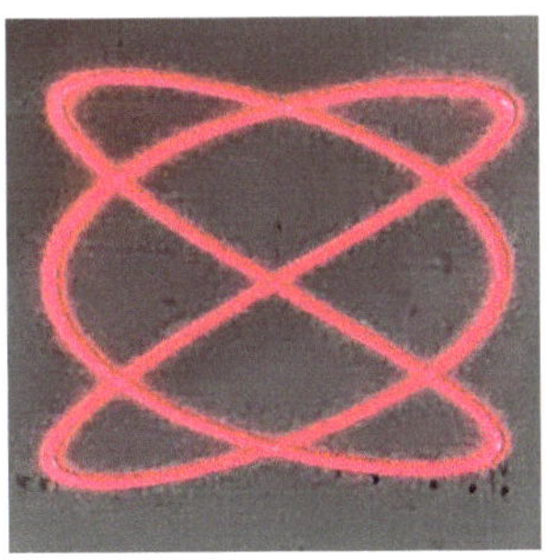

Figure 10. Lissajous scanning with the frequencies of the x and y scans being 70 and 58 Hz, respectively.

Lastly, an expanded laser beam is shined on a 1 × 3 scanner array (Device B ×2; Device C ×1). The three scanners are intentionally oriented along slightly different directions so that the reflected beams from the scanners can be separated. Figure 11a shows the reflected spots on a screen when no mirror is driven; in Figure 11b the traces of the reflected spots under a driving frequency of 217 Hz are presented. This type of scanner array can potentially be used for simultaneous multiple-beam steering.

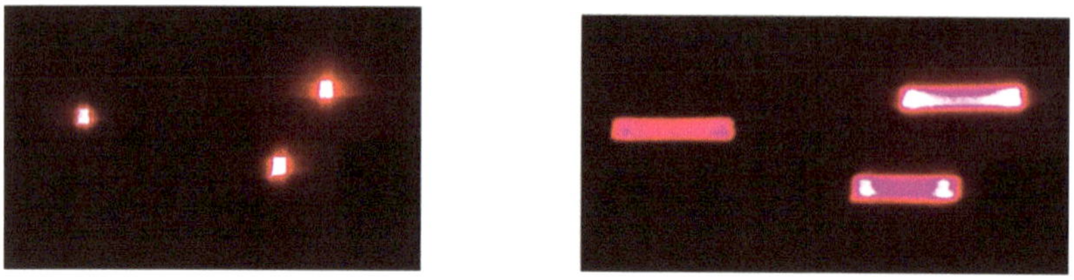

(a) (b)

Figure 11. (**a**) Reflected spots on a screen from a 1 × 3 scanner array when no mirror is driven; (**b**) Traces of the reflected spots under a driving frequency of 217 Hz.

5. Conclusions

In this paper, we have presented the designs and experimental results of 3D-printed 1-DOF optical scanners with large mirror areas (up to 20 × 20 mm^2). Five different designs have been studied. Each device consists of an aluminum-coated square silicon substrate serving as the mirror, two miniature permanent magnets, an electromagnet, and a 3D-printed structure including the mirror frame, torsion springs, and base. The scanner with a 20 × 20 mm^2 mirror area and a measured resonance frequency 84 Hz reaches a static half optical scan angle of 14.8 deg., i.e., a full optical scan angle of 29.6 deg., at 12 V. These scanners can be an extremely low-cost alternative to the conventional expensive galvanometer scanners. Although we have not achieved the 40 deg. seen in some galvanometer scanners, 29.6 deg. at a low voltage of 12 V and operation without needing an amplifier and closed-loop control are definitely attractive. Specifically, the cost of each scanner is <10 USD, which is an overwhelming margin against many competing devices. A short turnaround time of <1 week to complete fabrication and delivery is also appealing.

We have shown good angular repeatability of the scanner and also presented a Lissajous scanning pattern. A 1 × 3 scanner array has also been assembled, and multiple-beam steering implemented by this array has been demonstrated.

Author Contributions: Conceptualization, J.-c.T.; experiment, C.-K.S., Y.-N.H., G.-Y.L., W.-A.T., Y.-W.C., P.-H.Y.; writing, C.-K.S. and J.-c.T. All authors have read and agreed to the published version of the manuscript.

Funding: This research was supported by the Ministry of Science and Technology of Taiwan under Grant MOST 109-2221-E-002-019 and Grant MOST 110-2221-E-002-076.

Data Availability Statement: Data supporting the results reported in this paper may be obtained from the authors upon reasonable request.

Conflicts of Interest: The authors declare no conflict of interest.

References

1. Aylward, R.P. Advances and technologies of galvanometer-based optical scanners. In Proceedings of the SPIE'S International Symposium on Optical Science, Engineering, and Instrumentation, Denver, CO, USA, 18–23 July 1999.
2. Duma, V.-F. Laser scanners with oscillatory elements: Design and optimization of 1D and 2D scanning functions. *Appl. Math. Model.* **2018**, *67*, 456–476. [CrossRef]
3. Li, Y. Single-mirror beam steering system: Analysis and synthesis of high-order conic-section scan patterns. *Appl. Opt.* **2008**, *47*, 386–398. [CrossRef] [PubMed]
4. Novanta Photonics. 62xxK and 83xxK Series, Galvanometers. Available online: https://novantaphotonics.com/product/62xxk-and-83xxk-series-galvanometers/ (accessed on 22 June 2022).

5. Thorlabs. Galvanometers. Available online: https://www.thorlabs.com/navigation.cfm?guide_id=2269 (accessed on 22 June 2022).
6. Scanlab. Galvanometer Scanners. Available online: https://www.scanlab.de/en/products/galvanometer-scanners (accessed on 22 June 2022).
7. Salimi, M.H.; Villiger, M.; Tabatabaei, N. New Model for Understanding the Relationship between Tissue Composition and Photothermal Optical Coherence Tomography Signals. In *Proc. SPIE 11655, Label-Free Biomedical Imaging and Sensing (LBIS) 2021*; SPIE Proceedings: Bellingham, WA, USA, 2021. [CrossRef]
8. Luo, Y.; Arauz, L.J.; Castillo, J.E.; Barton, J.K.; Kostuk, R.K. Parallel optical coherence tomography system. *Appl. Opt.* **2007**, *46*, 8291–8297. [CrossRef] [PubMed]
9. Cui, M.; Lin, J.; Cheng, Z.; Gan, W. Jitter suppression for resonant galvo based high-throughput laser scanning systems. *Opt. Express* **2020**, *28*, 26414–26420. [CrossRef]
10. Li, Y.; Cui, T.; Li, Q.; Zhang, B.; Bai, Y.; Wang, C. A study of correction method to the pincushion distortion based on dual galvanometer LiDAR scanning system. *Optik* **2019**, *181*, 555–561. [CrossRef]
11. Wang, Q.; Wang, W.; Zhuang, X.; Zhou, C.; Fan, B. Development of an Electrostatic Comb-Driven MEMS Scanning Mirror for Two-Dimensional Raster Scanning. *Micromachines* **2021**, *12*, 378. [CrossRef] [PubMed]
12. Strathman, M.; Liu, Y.; Keeler, E.G.; Song, M.; Baran, U.; Xi, J.; Sun, M.T.; Wang, R.; Li, X.; Lin, L.Y. MEMS scanning micromirror for optical coherence tomography. *Biomed. Optics Express* **2015**, *6*, 211–224. [CrossRef] [PubMed]
13. Gorecki, C.; Bargiel, S. MEMS Scanning Mirrors for Optical Coherence Tomography. *Photonics* **2020**, *8*, 6. [CrossRef]
14. Yalcinkaya, A.D.; Urey, H.; Brown, D.; Montague, T.; Sprague, R. Two-Axis Electromagnetic Microscanner for High Resolution Displays. *J. Microelectromech. Syst.* **2006**, *15*, 786–794. [CrossRef]
15. Tanguy, Q.A.A.; Gaiffe, O.; Passilly, N.; Cote, J.-M.; Cabodevila, G.; Bargiel, S.; Lutz, P.; Xie, H.; Gorecki, C. Real-time Lissajous imaging with a low-voltage 2-axis MEMS scanner based on electrothermal actuation. *Opt. Express* **2020**, *28*, 8512–8527. [CrossRef] [PubMed]
16. Hashimoto, M.; Taguchi, Y. Design and Fabrication of a Kirigami-Inspired Electrothermal MEMS Scanner with Large Displacement. *Micromachines* **2020**, *11*, 362. [CrossRef] [PubMed]
17. Senger, F.; Albers, J.; Hofmann, U.; Piechotta, G.; Giese, T.; Heinrich, F.; von Wantoch, T.; Gu-Stoppel, S. A bi-axial vacuum-packaged piezoelectric MEMS mirror for smart headlights. In *MOEMS and Miniaturized Systems XIX*; SPIE Proceedings: Bellingham, WA, USA, 2020; Volume 1129305.
18. Kim, J.; Lee, H.; Kim, B.; Jeon, J.; Yoon, J.; Yoon, E. A high fill-factor micro-mirror stacked on a crossbar torsion spring for electrostatically-actuated two-axis operation in large-scale optical switch. In Proceedings of the Sixteenth Annual International Conference on Micro Electro Mechanical Systems, Kyoto, Japan, 23–23 January 2003. [CrossRef]
19. Conant, R.A.; Nee, J.T.; Lau, K.Y.; Muller, R.S. A flat high-frequency scanning micromirror. In Proceedings of the Technical Digest 2000 Solid-State Sensor & Actuator Workshop, Hilton Head Island, SC, USA, 4–8 June 2000; pp. 6–9.
20. Duma, V.-F.; Dimb, A.-L. Exact Scan Patterns of Rotational Risley Prisms Obtained with a Graphical Method: Multi-Parameter Analysis and Design. *Appl. Sci.* **2021**, *11*, 8451. [CrossRef]
21. Li, A.; Yi, W.; Zuo, Q.; Sun, W. Performance characterization of scanning beam steered by tilting double prisms. *Opt. Express* **2016**, *24*, 23543. [CrossRef]
22. Li, Y. Third-order theory of the Risley-prism-based beam steering system. *Appl. Opt.* **2011**, *50*, 679–686. [CrossRef] [PubMed]
23. Urey, H. Torsional MEMS scanner design for high-resolution scanning display systems. In *Optical Scanning*; SPIE Proceedings: Bellingham, WA, USA, 2002; Volume 4773, pp. 27–38. [CrossRef]
24. Farah, S.; Anderson, D.G.; Langer, R. Physical and mechanical properties of PLA, and their functions in widespread applications—A comprehensive review. *Adv. Drug Delivery Rev.* **2016**, *107*, 367–392. [CrossRef] [PubMed]
25. Hayakawa, T.; Watanabe, T.; Senoo, T.; Ishikawa, M. Gain-compensated sinusoidal scanning of a galvanometer mirror in proportional-integral-differential control using the pre-emphasis technique for motion-blur compensation. *Appl. Opt.* **2016**, *55*, 5640–5646. [CrossRef] [PubMed]
26. Carrasco-Zevallos, O.M.; Viehland, C.; Keller, B.; McNabb, R.P.; Kuo, A.N.; Izatt, J.A. Constant linear velocity spiral scanning for near video rate 4D OCT ophthalmic and surgical imaging with isotropic transverse sampling. *Biomed. Opt. Express* **2018**, *9*, 5052–5070. [CrossRef] [PubMed]
27. Hwang, K.; Seo, Y.-H.; Ahn, J.; Kim, P.; Jeong, K.-H. Frequency selection rule for high definition and high frame rate Lissajous scanning. *Sci. Rep.* **2017**, *7*, 14075. [CrossRef] [PubMed]

photonics

MDPI

Article

Deformation Measurement of a SS304 Stainless Steel Sheet Using Digital Image Correlation Method

Appurva Jain [1,*], Abhishek Mishra [1], Vikrant Tiwari [2], Gurminder Singh [3], Ravinder Pal Singh [4] and Sunpreet Singh [5,6]

1 Department of Mechanical Engineering, National Institute of Technology Delhi, New Delhi 110036, India
2 Department of Applied Mechanics, Indian Institute of Technology Delhi, New Delhi 110016, India
3 Department of Mechanical Engineering, Indian Institute of Technology Bombay, Mumbai 400076, India
4 Department of Mechanical Engineering, MMEC, Maharishi Markandeshwar (Deemed to be University), Mullana 133207, India
5 Department of Mechanical Engineering, National University of Singapore, Singapore 117549, Singapore
6 Department of Mechanical Engineering, Chandigarh University, Sahibzada Ajit Singh Nagar 140413, India
* Correspondence: appurva@nitdelhi.ac.in

Abstract: The digital image correlation (DIC) method is widely used in deformation measurements as it has the advantages of being a non-contact, high precision method that provides full field measurements, and requires simple experimental equipment. Traditionally, the grayscale speckle patterns captured by a monochromatic camera are used in the DIC method. With the growing development of consumer color cameras, there is great potential for developing color information in the DIC method. This paper proposes a displacement- and stress–strain-invariant DIC deformation measurement method based on the integer-pixel matching approach for speckle patterns during a tension test. For the integer-pixel matching stage, the load and displacement and stress–strain-invariant histories feature is used to estimate the initial value of the deformation parameters. In addition, this paper proposes a reverse retrieve strategy, instead of a forward search, to reduce the search time. Experiments show that the proposed DIC deformation measurement approach is not only capable of displacement invariance measurement, with robustness and high efficiency, but also that the average accuracy of the stress–strain result can reach 0.1%.

Keywords: digital image correlation; displacement measurement; numerical simulation; stress–strain comparison; speckle pattern

1. Introduction

Demand for lightweight parts, better product performance, efficiency, and higher safety is rising in the automotive sector [1–3]. Detailed measurements of crucial material parameters, such as strain limit, strength coefficients, and anisotropy coefficients, are necessary to optimize the design and production of these components [4–6]. Tension testing with an extensiometer is the technique most frequently used to determine a material's properties [7]. However, this system is not appropriate for post-diffuse necking and only provides an average strain over the specimen gauge length. A cutting-edge method for precise strain assessment is Digital Image Correlation (DIC) [8]. This method is ideal for characterizing material properties in the elastic and plastic ranges, since it allows for quick data collection. Additionally, it provides a noncontact method for carrying out complete-field, extremely high-precision measurement of displacement and strain. In the consideration of affordable and lightweight goods, the modern industrial sector needs materials with better strengths, such as SS304. To determine the material qualities needed for industrial applications, SS304 must be tested. For material processing, including stretching, stamping, bending, and other processes, tensile test data, such as the stress–strain relation, are essential [9–11]. The classic extensiometer method is typically used to calculate the average strain over a region by measuring the tensile strain between two sites. Using a tensile test machine with an extensiometer to determine the strain information at the necking point is nearly impossible. The entire measurement area between the extensiometerprobes is where the measured strain is spread out. During the tensile test, the DIC system measures the full-range strain distribution [8,12–14]. This study will examine field deformation using DIC strain measurement in tension testing, and contrast the results with the load vs. displacement and stress–strain relationship obtained using the conventional approach. Testing was conducted on SS304, which is an example of an advanced high-strength steel (AHSS) material. The measurement of SS304 is used in this paper as an example of how the test procedure works. Strain fields can be quantified using a variety of techniques, including photo elasticity, electronic speckle-pattern interferometry, and digital image correlation (DIC) [4,14–16].

The last 10 years have seen rapid progress in data processing and digital imaging technologies, which has increased interest in optical digital image techniques for strain assessment [8,13,14,17–20]. Strain fields of different sizes can be measured using the optical digital-image approach. Failure and damage can result from the localization of deformation within the tiny zones of the specimen. As a result, localization effects are crucial for comprehending material failure. To fully appreciate the localization impact, strain measurement and residual stress measurement appear to be crucial. There are several approaches for measuring residual stress, including X-ray diffraction, and the ultrasonic, eddy current, active magnetic, and passive magnetic methods [4,21]. The digital image correlation approach for measuring strain is the main topic of this study. In this study, a local deformation pattern is visualized using the digital image strain measuring approach, which is integrated with DIC. In recent years, DIC has been applied to numerous materials and mechanics research laboratories for full-field strain mapping applications. Theoretically, DIC can attain high accuracy and dependability. In reality, a number of other variables, including the deformation of a speckle on the specimen's surface, and the fixture method of the image capture instruments, might affect measurement precision during a mechanical test.

The range and frequency response of conventional contact measurement methods, such as mechanical extensometers and strain gauges, are limited, and thus do not give enough data to address the complexity of dynamic mechanical behavior. To quantify stress and deformation fields in tests, non-contact full-field approaches, such as photoelasticity, Moiré, caustics, coherent gradient sensors, and digital image correlation (DIC), have significant advantages [22]. The DIC technique, first proposed by [23], is a promising tool for ductile heterogeneous materials across a wide range of length and time scales [24–32]. This is due to the recent advancement of image processing methods and the introduction of CCD (charge-coupled device) cameras. On the one hand, the method is applicable to studies encompassing a wide range of fields of view, from the nano-/micro-scale to the field-scale, as well as from two-dimensional (2D) to three-dimensional, because it has no inherent length scale (3D). On the other hand, the DIC methodology has also been tested over a wide variety of loading rates, leading to improvements in the advent of CCD-cameras with high spatial and temporal resolutions, modern image processing algorithms, and numerical computations. Assessing the performance of the DIC approach for materials testing is one of the goals of the current paper.

The structure of this paper is as follows. In order to confirm the accuracy of the strain estimation utilizing the load vs. displacement obtained through the DIC technique, a direct comparison experiment of the elastic strain measurement is first offered. Then, in order to observe the strain distribution during the test, the application of the DIC approach to assess the nonhomogeneous deformation of an SS304 rectangular sample during a tensile test is shown.

2. DIC Methodology

DIC is a non-contact 3D full-field optical method for detecting contours, deformation, vibration, and strain on practically any material. The method is applicable to a wide range of static and dynamic tests, including tensile, torsion, bending, and combination loads. DIC is less costly and simpler than other techniques, such as speckle interferometry, and it is more accurate and subjective than manual measuring methods, offering up a wide range of possible applications. To monitor a group of pixels (called subsets) in the distorted and reference pictures, as shown in Figure 1, DIC measurement uses temporal matching and correlation functions.

A charge-coupled device (CCD) camera collects picture data, which is then converted into digital form and stored in a computer for analysis. The DIC technique includes tracking the position of many surface points in two consecutive images using a correlation algorithm to determine displacement information. Displacements must be interpolated as weight functions in finite element form in order to calculate stress and strains using the correlation function. The correlation algorithm is based on tracking the intensity (Gray value) pattern in discrete subsets of neighborhoods during movement (shown in the area with a dashed line; one pane represents an image pixel), as shown in Figure 1. By contrasting the two picture subsets, the correlation algorithm calculates the local displacement values, U and V. A (n × m) pixel area that is rectangular in shape defines the area of interest in the reference image. The appropriate subset, which is likewise a (n × m) pixel region, is estimated at a specific point with a certain range in the distorted image. The range in the deformed image is searched for carefully, pixel by pixel.

By entwining the subset from the deformed image with the broader subset from the reference image, the algorithm produces the cross-correlation factor, C (Equation (1)) for a domain of theoretical displacements, (U, V), in 1-pixel increments as described in:

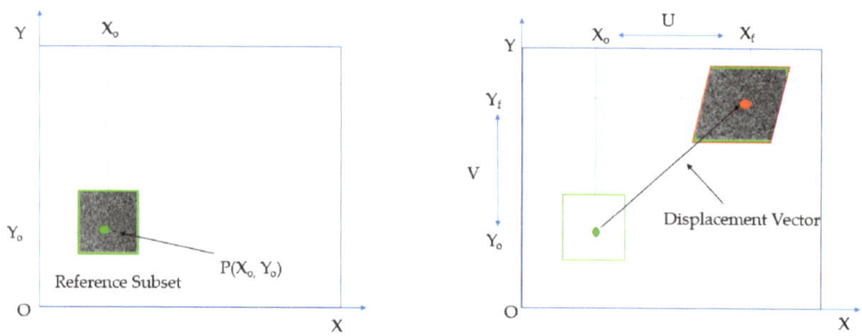

Figure 1. Temporal matching for the correlation function generation of unreformed and deformed images.

Stress and Strain Calculation from the Correlation Function

DIC measurement uses temporal matching and correlation functions, shown in Equation (1).

The working procedure is explained as follows:

- The digital image is first separated into smaller parts known as subsets.
- Subset shape functions are imposed on the reference subset to account for the deformed shape of the subset in the deformed picture.
- The distorted position of the subset may not be at the integer location.
- The correlation function (C) is constructed to match the similarity of the subset in the un-deformed and deformed images.

$$C = \frac{\oint_{Am}^{0}(F(x,y)(G(X_0+U,Y_0+V))dA}{\left[\int_{AM}^{0}[(F(x,y)]^2 \int_{Am}^{0}[(G(X_0+U,Y_0+V))]^2 dA\right]^{\frac{1}{2}}} \quad (1)$$

where G is the severity of the pixels in the reference subset, dA is the severity of the pixels in the deformed subset, and (X_0, Y_0) are positional subset coordinate axes. Its origin is at the center of the subset at the control point.

In this paper, the Newton–Raphson method is used to estimate strain and stress. If we want to calculate the strains at the current position, we first choose a square window that contains discrete points that are (2m + 1) by (2m + 1) all around it. This window is known as the strain calculation window. The displacement distributions within it can be roughly represented as a linear plane if the strain calculation window is small enough.

Thus, we have the following equation:

$$\begin{aligned} U(i,j) &= a_0 + a_1 x + a_2 y \\ V(i,j) &= b_0 + b_1 x + b_2 y \end{aligned} \tag{2}$$

where *a*,*b* is the unknown polynomial coefficient that determines the displacement relationship with the m coordinate frame. *U*, *V* are the reference displacement at location (*i*,*j*) obtained by DIC, as indicated in Figure 1. Equation (2) can be rewritten in matrix form:

$$\begin{bmatrix} 1 & m & m \\ 1 & m-1 & m \\ \vdots & \vdots & \vdots \\ 1 & 0 & 0 \\ \vdots & \vdots & \vdots \\ 1 & -m+1 & -m \\ 1 & -m & -m \end{bmatrix} \begin{pmatrix} a_0 \\ a_1 \\ a_2 \end{pmatrix} = \begin{pmatrix} U(m.m) \\ U(m-1,m) \\ \vdots \\ U(0,0) \\ \vdots \\ U(-m,-m) \\ U(-m+1,-m) \end{pmatrix} \tag{3}$$

Therefore, it is possible to solve the unknown polynomial coefficients using the Newton–Raphson method. It is vital to remember that the strain calculation window may contain fewer points than (2m + 1) (2m + 1) for points at the image boundary or in the region of the discontinuity area. However, by ignoring these faulty points within the local strain calculation window from the above equation, we can still compute the strain components using the Newton–Raphson iteration method.

$$\begin{aligned} u_i &= u_0 + \tfrac{\partial u}{\partial x}\Delta x_i + \tfrac{\partial u}{\partial y}\Delta y_i \\ v_i &= v_0 + \tfrac{\partial v}{\partial x}\Delta x_i + \tfrac{\partial v}{\partial y}\Delta y_i \end{aligned}$$

After obtaining the displacements and displacement gradients at the point (*x*,*y*), the full field strain can be calculated under the small deformation assumption:

$$\varepsilon_x = \frac{\partial u}{\partial x} \tag{4}$$

$$\varepsilon_y = \frac{\partial v}{\partial y} \tag{5}$$

$$\gamma_{xy} = \frac{\partial u}{\partial y} + \frac{\partial v}{\partial x} \tag{6}$$

The unknown parameters $\left(u, v, \frac{\partial u}{\partial x}, \frac{\partial u}{\partial y}, \frac{\partial v}{\partial x}, \frac{\partial v}{\partial y}\right)$ are determined by using the Newton–Raphson method to minimize the correlation function.

3. Experiments

The experiments were performed on a universal testing machine (Zwick Roell 250 with a maximum capacity of 250 kN, as shown in Figure 2b) with homogenous working conditions at room temperature. The strain rate was kept $\dot{\varepsilon} = 1 \times 10^{-3}$ per second. The material used for the tensile test was SS304 with 1 mm thickness. A series of samples were cut in a 0° rolling direction. Rectangular cross sections with dimension 200 mm × 20 mm × 1 mm were cut out from the steel sheet. While performing the experiments, 50 mm from both ends was constrained, and the remaining mid-section (100 mm in length) was kept under loading conditions. The samples' length, width, and thickness were measured along the *y*-axis, the *x*-axis, and the *z*-axis, respectively. The load was applied in the y-direction along the length. The tests were conducted for the samples in the rolling direction until three reproducible curves were produced.

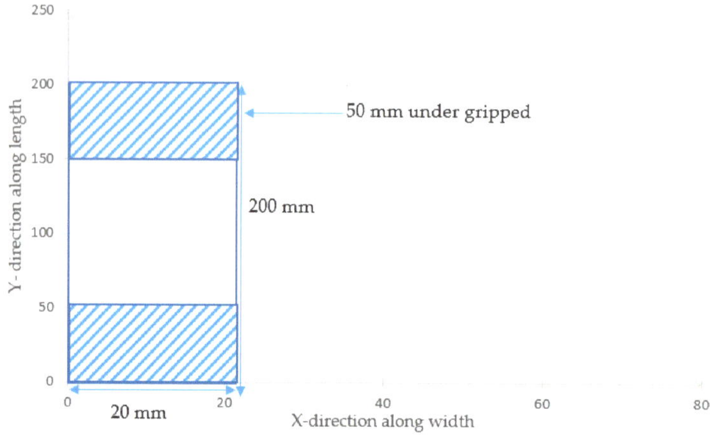

(a)

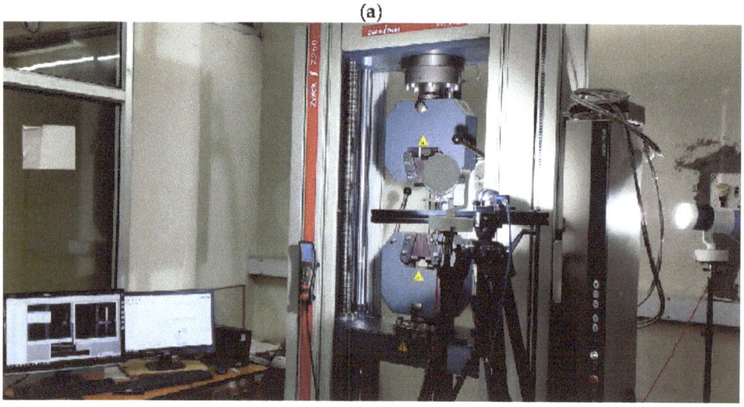

(b)

Figure 2. (**a**) Test specimen with dimensions and a coordinate frame. (**b**) Experimental setup for the tension test with the DIC devices.

The test specimen with dimensions and a coordinate frame is shown in Figure 2a. The strain was measured using the DIC device and a POINTGREY® camera. The DIC system is helpful for measuring strain distribution in the plastic zone of the sample. The Cauchy stress vs. logarithmic strain curves were computed using results from the DIC system that were acquired after post-processing. The strain was measured using a digital image correlation device (DIC). This guarantees that the results are accurate and removes the impact of any impurities or porosity in the material from the results.

The contour plot image was obtained from the DIC Vic-3D software, which was applied under tensile load conditions. Figure 3 shows the contour plot of the measured strain under tensile loading, which provides some parameters such as analysis types, the DIC radius, step analysis, image correspondence, units for pixel, and the correlation coefficient. Figure 3a shows the contour plot of the strain value on the stainless steel material, which recorded the development trend of the strain value. The necking area was detected and observed using the experimental DIC method, as shown in Figure 3b. Moreover, the maximum strain value red colour was 35 mm and minimum strain value blue colour was around 0 mm.

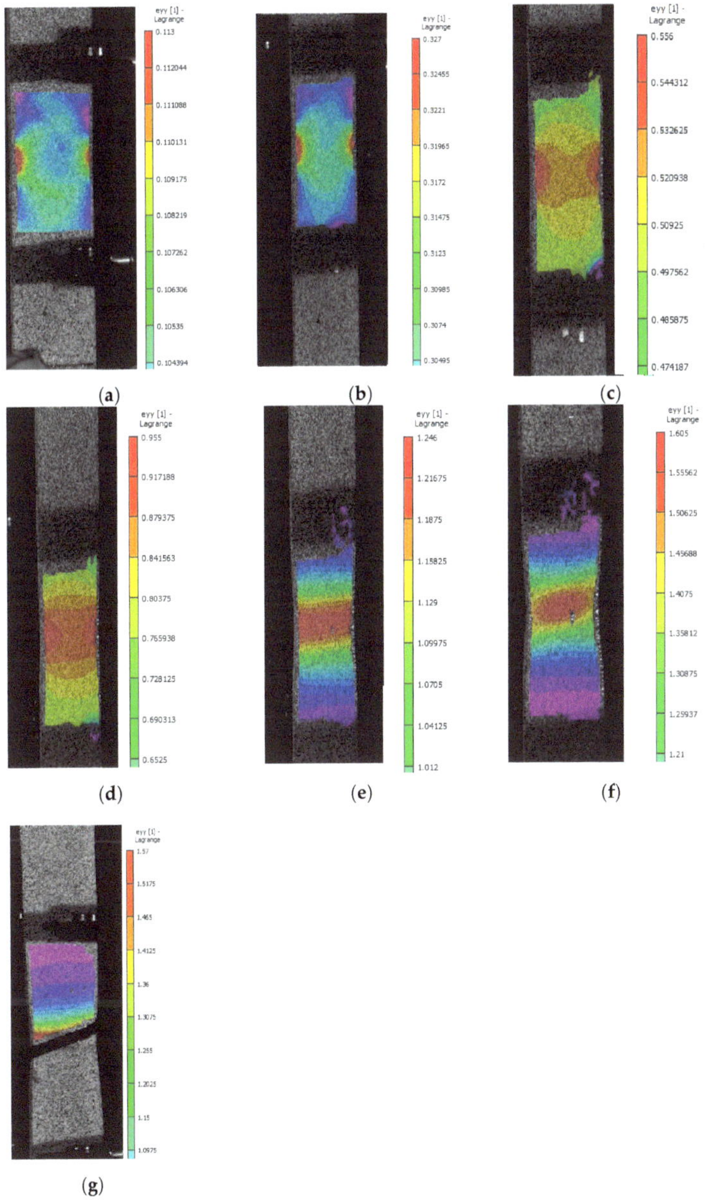

Figure 3. Strain contour plot of an SS304 steel sheet at the ultimate load with different strain legends. Figure 3 represent contour of test sample obtained from DIC. (**a**) ε_{yy} distribution at time 0 s. (**b**) ε_{yy} distribution at time 30 s. (**c**) ε_{yy} distribution at time 60 s. (**d**) ε_{yy} distribution at time 90 s. (**e**) ε_{yy} distribution at time 120 s. (**f**) ε_{yy} distribution at time 150 s. (**g**) ε_{yy} distribution at time 180 s.

4. Results

4.1. Experimental Results and Discussion

The stress–strain curve is plotted using data obtained from the DIC for the tension test samples. The plot for the 0° rolling direction is shown in Figure 4. While Cauchy stress vs. logarithmic strain curves are plotted up to the necking, nominal stress vs. logarithmic strain curves are plotted up to the rupture. The data are unreliable because, after necking, the nominal stress in the stress–strain curves starts to decrease. The outcomes show that the curves can be replicated. Reproducible results were obtained from similar tests carried out on the tensile test materials in each orientation. Figure 3b shows one from each orientation of the nominal stress–logarithmic-strain curve and the Cauchy stress–logarithmic-strain curve. This demonstrates that there is no discernible difference in the highest value of nominal stress and between curves from all orientations.

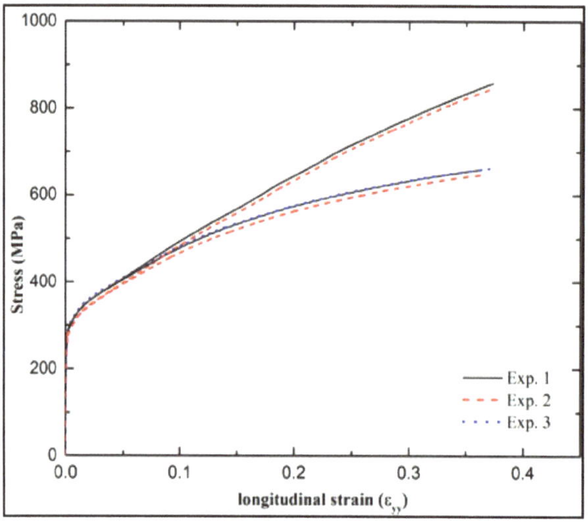

Figure 4. Nominal and Cauchy stress vs. logarithmic strain for reproducible experiments.

4.2. Numerical Results and Discussion

Using an elasto-plastic model with homogeneous material properties, the tensile test sample was numerically simulated. The material's mechanical properties were utilized in accordance with the experimental sections stated above. As shown in Figure 5, the simulation with symmetry boundary conditions was run as per the eighth sample along the x, y, and z directions. The sample's 3D model was created using the following dimensions: 50 mm × 10 mm × 0.5 mm. As shown in Figure 5, a 5 mm displacement was applied to the model's top surface, along the y-axis. The commercial finite element code ABAQUS/Standard was used for the simulation. Only the area subjected to the load was taken into consideration for the numerical simulation of the sample, disregarding the area beneath the grips during the experiments (50 mm at both ends of the sample). Calculations of various stresses and strains were made using the simulation. The stress–strain data were post-processed using several macroscopic criteria to compute the damage factor, as mentioned in the preceding sections. The C3DR element was used to generate the mesh. The linear mesh was defined as 0.2 mm to 20 mm in length, away from the sample's center (half-length under the extensometer, as the symmetric sample is considered). After that, a biased mesh was taken along the length, ranging in size from 0.2 to 0.5 mm. A linear mesh with a 0.2 mm size was taken across the width. Young's modulus and the hardening curve were taken into consideration as typical properties when simulating the sample at 0°/RD.

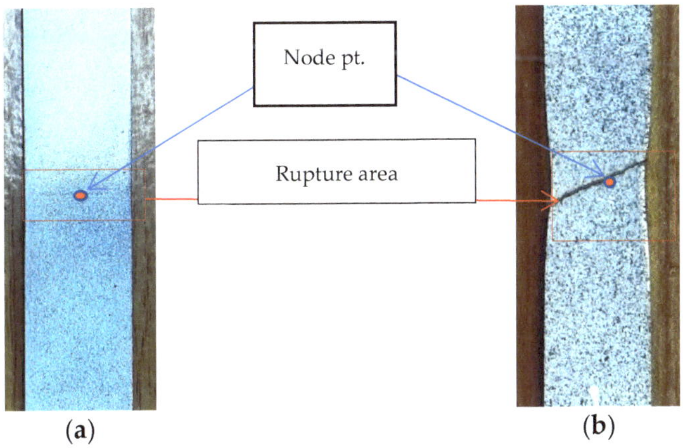

Figure 5. Tension test samples; (**a**) original sample and (**b**) sample after rupture.

The simulation result was considered up to the point when the extensiometer displacement in the simulation reached the same value as in the experiment (i.e., the rupture point). The various stress–strain data obtained from the numerical simulation of the tensile test sample are presented in Figure 6. They indicate the distribution of (a) von-Mises stress, (b) maximum principal stress, (c) equivalent plastic strain, and (d) hydrostatic stress at the point of rupture. The load–extensiometer-displacement curve is presented in Figure 7. It indicates that there is good agreement between the experimental curve and the simulated curve. The maximum load value reached in experiment was 13,702 N and in the simulation it was 13,331 N, whereas the extensiometer displacement at the rupture point was 27 mm in both the experiment and the simulation.

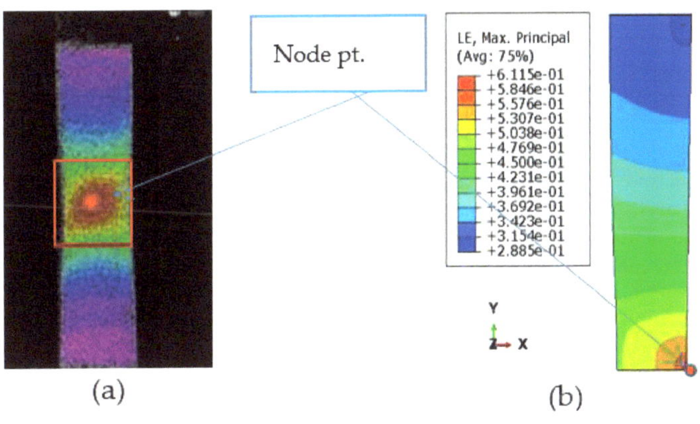

Figure 6. Tensile test: images of the specimen; (**a**) distribution of strain along length (via digital image correlation (DIC)) and (**b**) logarithmic strain distribution just before rupture.

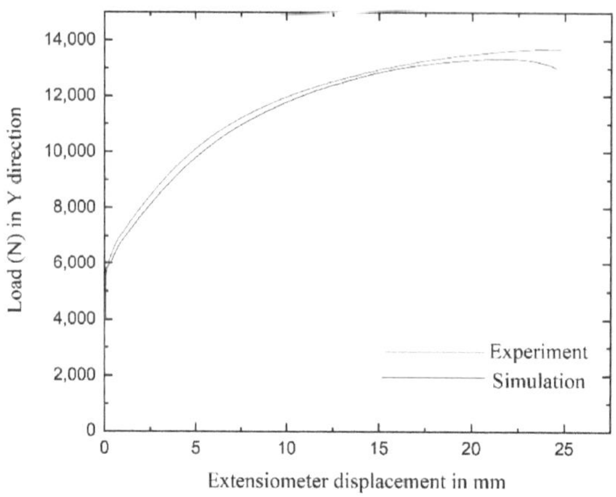

Figure 7. Load–Extensiometer-displacement curves for the simulation and for the experiment.

5. Validation

A reverse retrieve strategy was used to validate the proposed deformation measurement approach. In this strategy, a pixel of speckle samples was used to evaluate deformation by node-point matching during the simulation. To verify the feasibility of this matching technique, it was further compared with experiment results. Moreover, traditional equations were used to obtain stress and strain data for the experiment and for the numerical simulation.

5.1. Simulation

The first validation concerns the displacement invariance of the samples, whereby a speckle pattern was used for verification, as shown in Figure 5. Figure 5a was displaced by 0 and Figure 5b was obtained after rupture. To better conform the actual measurement situation, the load in the experiments was homogenously added to these samples. Subsequently, in the numerical simulation, a node point at the highest strain deformation was uniformly selected in Figure 6a,b, and this node point was matched in the tension test experiment using the DIC device. Since the deformation is known from the true displacement of the generated samples, the true displacement of these node points is known. The performance of the DIC deformation measurement technique was evaluated by the error between the simulated results and the true displacements, as shown in Figure 6. On samples with node points, it can be demonstrated that the DIC deformation measurement technique can match displacement with a maximum inaccuracy of roughly 3%. The precision of the DIC deformation measurement technique fully satisfies the experimental value's accuracy criterion, as the convergence range of the simulation approach is roughly 7% [16].

The validation of the stress–strain curve obtained from the tension test experiment is concerned with the numerical simulation in same rolling direction. The simulation was run in the same working conditions as were employed in the experiments. In the simulation, the calculation was performed by taking the centroidal node element and extracting the parameters of Cauchy stress and logarithmic strain value; these values were further compared with the DIC-obtained data. The experimentally obtained load value needed to be converted into Cauchy stress, as DIC gives logarithmic strain values.

5.2. Experimental

Traditional calculations were performed for the deformation measurement. Stress and strain are vital parameters which indicate the deformation of any material. The calculation was conducted by continuously measuring gauge length (l_0) (i.e., the calibrated distance between two marked surfaces of a specimen) until rupture. The strain yield leads to the measurement of sample deformation. This is then calculated using the ratio of the increase in the specimen gauge's length to its initial gauge length, represented in Equation (7):

$$\delta = \frac{(L - L_0)}{L_0} \tag{7}$$

Tensile stress σ is calculated as the ratio of the tensile load (F) that was applied to it and the specimen's initial cross-section area (A_0), represented in Equation (8):

$$\sigma = \frac{F}{A_0} \tag{8}$$

As DIC measures strain at every instant, Cauchy stress and logarithmic strain were obtained for the test sample in the 0° rolling direction. Relative elongation and relative reduction of area are obtained as parameters of the deformation measurement of the material.

Relative elongation is represented by Equation (9):

$$\delta = \frac{(L_m - l_0)}{L_0} \tag{9}$$

where (L_m) is the maximum length of the specimen.

Relative reduction in area is the ratio between the decrease in the area of the specimen's cross section before its rupture and its original cross-sectional area, represented by Equation (10):

$$\psi = \frac{(A_0 - A_{min})}{A_0} \tag{10}$$

where (A_{min}) is the minimum specimen cross-sectional area.

The comparison of the numerical simulation and the experimental values was plotted into a stress–strain curve, which shows an error of roughly 0.1% (see Figure 7).

5.3. Reverse Retrieve Strategy

The numerical simulation and experimental displacement of the DIC measurement uses color histogram features, which have the advantages of scaling and rotation invariance. The reverse retrieve strategy can also be used to improve the efficiency of the DIC deformation technique. In addition, the stress–strain comparison makes the efficacy of DIC measurement technique more accurate. Therefore, in the experimental section, simulation experiments were conducted for the evaluation of the displacement invariance, deformation invariance, and computational efficiency of the DIC measurement technique. It can be seen from the experimental results that the reverse retrieve strategy can also be used for validation, and meets the accuracy requirements of deformation measurement methods. Furthermore, the search strategy based on reverse retrieval can greatly improve the matching speed, which is faster than the state-of-the-art DIC displacement measurement method. In actual experiments, the test specimen was stretched and measured using an extensiometer and tension test equipment. The deformation of the test specimen was measured using the DIC deformation measurement technique, and the error was estimated to be around 0.1%, which satisfies the majority of measurement criteria. The suggested approach focuses on addressing the measurement of sample deformation during mechanical testing. At this time, color paint must be properly sprayed in a speckle pattern, rather than as a naturally occurring surface. To solve this issue, it is anticipated that feature acquisition and matching methods will be improved.

When using the above strategy, the computational burden is greatly increased due to the use of pixel-to-node point matching in the simulation. The different scales need to be extracted and matched. For instance, the area of the subset of interest is 30 pixels for obtaining the highest strain value at the rupture point, and then the simulation contains a total of 390 nodes, which means that the computational burden of obtaining the highest strain value is increased by 13 times. Moreover, when the range of the scaling factors is unclear, the computational burden will be much greater. To improve efficiency, this paper proposes a search strategy based on the reverse retrieve method.

The traditional method adopts a forward strategy; that is, the reference sub-region is searched pixel by pixel in all deformed images to find the best matching position, as shown in Figure 8. With the rupture of the material, the deformation feature of forward matching will also change. Once the feature is changed, it is necessary to perform feature extraction on all nodes in the new simulation, which is very time consuming. To avoid redundant calculations, a reverse retrieve strategy is proposed, in which some deformed subregions are selected and matched with the reference image, as shown in Figure 8. By using this method, the search space is the reference image, i.e., before loading, which is constant and can be precalculated. The reverse search strategy only needs to perform feature extraction on the search space once; therefore, it is more efficient than the forward search strategy.

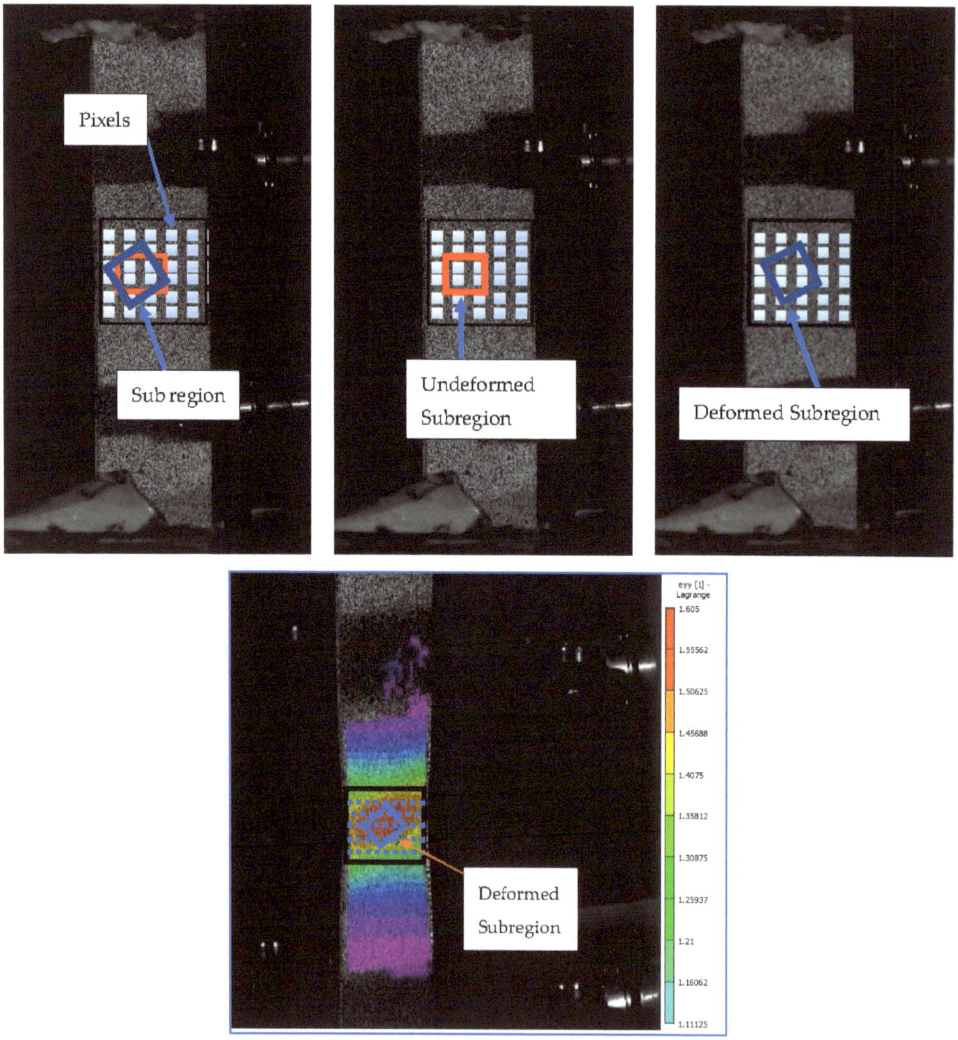

Figure 8. The figures show that the reverse search method searches the deformed sub-regions (the red and blue boxes) in the deformed image, which is constant and precalculated.

Before the matching, the displacement feature of the multiscale templates in the reference image are extracted pixel by pixel and stored in a feature set Q, which is constant and does not need to be updated. When the deformed images are updated, only the displacement features of the new deformed subregion need to be extracted, and then they are retrieved in feature set Q. The value of the displacement is quantized and the retrieval-method-based search space is used to speed up image retrieval. Finally, the reverse retrieval strategy avoids the repeated feature extraction of the search space and greatly improves research efficiency.

The figures show that the reverse search method searches the deformed sub-regions (the yellow and green boxes) in the reference image, which is constant and precalculated. For clarity, only two deformed images are displayed.

6. Conclusions

The speckle-pattern matching information has brought great advantages to the fields of computer vision and image retrieval. However, it has not been fully utilized in relation to the DIC method. To exploit the additional pattern information provided by color speckle patterns, this paper improves the DIC deformation measurement in two respects: displacement measurement and stress–strain curve matching. Specifically, this paper proposes that the performance of the DIC deformation measurement method can be enhanced by: (i) load vs displacement measurements based on integer-pixel matching; and (ii) a comparison based on the stress–strain histories strategy. For validation, experiments with simulated color images were performed, and show that our method have the advantages of measurement in displacement invariance, high efficiency, and deformation invariance. For the property measurement of the real material, uniaxial tension tests were conducted, where the average error of the strain results reached 0.1%.

The traditional extensiometer method for tensile tests can only provide a stress–strain curve before the tensile load reaches its peak value or before substantial strain localization occurs. An important and unique feature of the DIC-assisted tensile test is that it can provide a more complete true stress–strain curve after the tensile load passes its peak and drops into a "necking" stage, until the specimen's complete separation. This material mechanical response information is critical for numerical simulations that correctly visualize the formation history.

A comparison of the DIC technique with the traditional extensiometer-based technique is presented in Tables 1 and 2.

Table 1. Comparison of the strain measurements produced by different mechanisms and their possible applications.

Type	Approach	Mechanism of Mechanical Strain Evaluation	Test	Image Analysis	Advantage	Disadvantage
Electro-based	(1) Strain gauge [28] (2) Strain transducers [29]	The deformation of the material induces the electrical signal changes, which can be converted into the strain values of the materials.	Discrete In vivo Ex vivo	N/A	Cheap; offline work	Invasive; low anti-interference
Light -based	(1) Microscopy camera [30]	The relative strain is assessed by comparing the images before and after the material deformation.	Serial Ex vivo	Marker-tracking algorithm	Cheap; easy operation	Transparent or translucent samples
	(2) Photo-elastic [15]	Strain measurement based on *load-induced (stress-induced) birefringence.*		Cheap; easy application	time-consuming	Regional phase unwrapping algorithm

Table 1. *Cont.*

Texture-correlation-based	(1) Speckle interferometry [31]	Strain is quantified from changed search space patterns in the images during the deformation of the materials.	Serial In vivo Ex vivo	Simple structure	Complex analysis	Baseband speckle tracking algorithm; Registration algorithm
Imaging-based	(1) Moiré [22]	The Moiré effect is the mechanical interference of light by a superimposed network of lines.	In-plane fringes; out-plane fringes	Fast; non-invasive; portable; cost-efficient and has no harmful radiation	Resolution; costly	Moiré fringe phase shifting measurement algorithm
	(2) DIC [16]	Strain is evaluated by tracking the subsets including markers or speckles on the surface of tissues.	Serial Ex vivo	Relatively fast imaging; relatively low cost	Expensive; high contrast	Computed tomography
	(3) DVC [32]	Strain is evaluated by tracking image subsets by tracking the natural pattern in the tissues.	Serial In vivo Ex vivo			Correlation-based algorithm

Table 2. Comparison between DIC system and Extensiometer.

DIC System	Externsiometer
Non-contact measurement	Contact measurement
Unlimited number of deformation measurements	An extensometer can be used only once (a glued extensometer cannot be peeled off without damaging it)
The possibility of testing samples of any shape and material; the tested surface of the sample does not have to be flat	A surface on which the extensometer is glued has to be flat
The ability to measure deformation in all directions (along the X-axis, Y-axis, and Z-axis), on a plane or in three-dimensional space	The ability to measure deformation only in the chosen direction
Full-field deformation analysis	Results of the deformation at selected points of the sample, i.e., at the points where the sensors are attached
A measurement of the real maximum displacements and deformations	A measurement limited by the maximum value of the deformation of an extensometer
A quick preparation of a random pattern of black dots on a white background on the sample surface by spraying paint	A time-consuming process of placing the extensometer on the surface of a sample (gluing, etc.)
The need to clean the surface of a sample before testing	

Author Contributions: Conceptualization, A.J., V.T. and G.S.; methodology, R.P.S., A.J. and S.S. validation, A.J., A.M., V.T. and G.S. data curation, A.J., A.M., S.S. and R.P.S., writing—original draft preparation, A.J., A.M. and V.T., writing—review and editing, A.J., G.S. and S.S. All authors have read and agreed to the published version of the manuscript.

Funding: This research received no external funding.

Data Availability Statement: Not applicable.

Conflicts of Interest: The authors declare no conflict of interest.

References

1. Pandre, S.; Morchhale, A.; Kotkunde, N.; Singh, S.K. Influence of Processing Temperature on Formability of Thin-Rolled DP590 Steel Sheet. *Mater. Manuf. Process.* **2020**, *35*, 901–909. [CrossRef]
2. Harikrishna, C.; Nagaraju, C. Modeling of Cylindrical Upsetting Process for Enhanced Ductile Fracture. *Mater. Today Proc.* **2020**, *39*, 1629–1634. [CrossRef]
3. Habibi, N.; Zarei-Hanzaki, A.; Abedi, H.R. An Investigation into the Fracture Mechanisms of Twinning-Induced-Plasticity Steel Sheets under Various Strain Paths. *J. Mater. Process. Technol.* **2015**, *224*, 102–116. [CrossRef]
4. Mishra, A.; Leguen, C.; Thuillier, S.; Maire, E. Investigation of Ductile Damage in DP980 Steel Sheets Using Mechanical Tests and X-ray Micro-Tomography. *AIP Conf. Proc.* **2011**, *1353*, 1464–1469. [CrossRef]
5. Kim, J.H.; Serpantié, A.; Barlat, F.; Pierron, F.; Lee, M.G. Characterization of the Post-Necking Strain Hardening Behavior Using the Virtual Fields Method. *Int. J. Solids Struct.* **2013**, *50*, 3829–3842. [CrossRef]
6. Jain, A.; Mishra, A. Comparative Study of Macroscopic Rupture Criteria. *Mater. Today Proc.* **2019**, *18*, 3394–3400. [CrossRef]
7. Chen, Y.; Clausen, A.H.; Hopperstad, O.S.; Langseth, M. Stress-Strain Behaviour of Aluminium Alloys at a Wide Range of Strain Rates. *Int. J. Solids Struct.* **2009**, *46*, 3825–3835. [CrossRef]
8. Gharehbaghi, S.; Gandomi, M.; Plevris, V.; Gandomi, A.H.; Abdulameer Kadhim, A.; Mohammed Kadhim, H.; Ham, S.W.; Cho, J.U.; Cheon, S.S.; Paresi, P.R.; et al. Fracture Prediction in Plastic Deformation Processes. *Ductile Fract. Met. Form.* **2020**, *7*, 1–17.
9. Jain, M.; Lloyd, D.J.; Macewen, S.R. Hardening Laws, Surface Roughness and Biaxial Tensile Limit Strains of Sheet Aluminium Alloys. *Int. J. Mech. Sci.* **1996**, *38*, 219–232. [CrossRef]
10. Gia Hai, V.; Thi Hong Minh, N.; Nguyen, D.T. A Study on Experiment and Simulation to Predict the Spring-Back of SS400 Steel Sheet in Large Radius of V-Bending Process. *Mater. Res. Express* **2020**, *7*, 016562. [CrossRef]
11. Lou, Y.; Yoon, J.W. A User-Friendly Anisotropic Ductile Fracture Criterion for Sheet Metal under Proportional Loading. *Int. J. Solids Struct.* **2021**, *217*, 48–59. [CrossRef]
12. Boyce, B.L.; Reu, P.L.; Robino, C.V. The Constitutive Behavior of Laser Welds in 304L Stainless Steel Determined by Digital Image Correlation. *Metall. Mater. Trans. A Phys. Metall. Mater. Sci.* **2006**, *37*, 2481–2492. [CrossRef]
13. Khare, S.; Kumar, K.; Choudhary, S.; Singh, P.K.; Verma, R.K.; Mahajan, P. Determination of Johnson–Cook Material Parameters for Armour Plate Using DIC and FEM. *Met. Mater. Int.* **2021**, *27*, 4984–4995. [CrossRef]
14. Jang, I.; Bae, G.; Song, J.; Kim, H.; Park, N. Fracture Envelopes on the 3D-DIC and Hybrid Inverse Methods Considering Loading History. *Mater. Des.* **2020**, *194*, 108934. [CrossRef]
15. Iosipescu, N. New Accurate Procedure for Single Shear Testing of Metals. *J. Mater.* **1967**, *2*, 537–566.
16. Jain, A.; Mishra, A.; Tiwari, V. Investigation of Rupture in SS304 Steel Sheet Using Macroscopic Criteria. *IOP Conf. Ser. Mater. Sci. Eng.* **2020**, *998*, 012015. [CrossRef]
17. Wang, L.; Park, J.H.; Choi, N.S. Observation of Notch Effect in Al6061-T6 Specimens under Tensile Loading Using Digital Image Correlation and Finite Element Method. *J. Mech. Sci. Technol.* **2020**, *34*, 1049–1058. [CrossRef]
18. Gardner, K.A. Experimental Techniques for Shear Testing of Thin Sheet Metals and Compression Testing at Intermediate Strain Rates. Master's Thesis, The Ohio State University, Columbus, OH, USA, 2013; p. 133.
19. Paul, S.K.; Roy, S.; Sivaprasad, S.; Bar, H.N.; Tarafder, S. Identification of Post-Necking Tensile Stress–Strain Behavior of Steel Sheet: An Experimental Investigation Using Digital Image Correlation Technique. *J. Mater. Eng. Perform.* **2018**, *27*, 5736–5743. [CrossRef]
20. Versaillot, P.D.; Wu, Y.F.; Zhao, Z.L. An Investigation into the Phenomenon of Macroscopic Plastic Deformation Localization in Metals. *J. Phys. Conf. Ser.* **2021**, *1777*, 012067. [CrossRef]
21. Pham, C.H.; Thuillier, S.; Manach, P.Y. Mechanical Properties Involved in the Micro-Forming of Ultra-Thin Stainless Steel Sheets. *Metall. Mater. Trans. A Phys. Metall. Mater. Sci.* **2015**, *46*, 3502–3515. [CrossRef]
22. Field, J.E.; Walley, S.M.; Proud, W.G.; Goldrein, H.T.; Siviour, C.R. Review of Experimental Techniques for High Rate Deformation and Shock Studies. *Int. J. Impact Eng.* **2004**, *30*, 725. [CrossRef]
23. Peters, W.H.; Ranson, W.F. Digital Imaging Techniques In Experimental Stress Analysis. *Opt. Eng.* **1982**, *21*, 213427. [CrossRef]
24. Hild, F.; Roux, S. Digital Image Correlation: From Displacement Measurement to Identification of Elastic Properties—A Review. *Strain* **2006**, *42*, 69–80. [CrossRef]
25. Tiwari, V.; Sutton, M.A.; McNeill, S.R. Assessment of High Speed Imaging Systems for 2D and 3D Deformation Measurements: Methodology Development and Validation. *Exp. Mech.* **2007**, *47*, 561–579. [CrossRef]
26. Bornert, M.; Brémand, F.; Doumalin, P.; Dupré, J.C.; Fazzini, M.; Grédiac, M.; Hild, F.; Mistou, S.; Molimard, J.; Orteu, J.J.; et al. Assessment of Digital Image Correlation Measurement Errors: Methodology and Results. *Exp. Mech.* **2009**, *49*, 353–370. [CrossRef]
27. Schreier, H.; Orteu, J.J.; Sutton, M.A. Image Correlation for Shape, Motion and Deformation Measurements: Basic Concepts, Theory and Applications. In *Image Correlation for Shape, Motion and Deformation Measurements*; Springer: Berlin/Heidelberg, Germany, 2009; 321p. [CrossRef]
28. Motra, H.B.; Hildebrand, J.; Dimmig-Osburg, A. Assessment of Strain Measurement Techniques to Characterise Mechanical Properties of Structural Steel. *Eng. Sci. Technol. Int. J.* **2014**, *17*, 260–269. [CrossRef]
29. Jia, B.; Rusinek, A.; Pesci, R.; Bernier, R.; Bahi, S.; Wood, P. A Novel Technique for Dynamic Shear Testing of Bulk Metals with Application to 304 Austenitic Stainless Steel. *Int. J. Solids Struct.* **2020**, *204*, 153–171. [CrossRef]

30. Kajberg, J.; Sundin, K.G.; Melin, L.G.; Ståhle, P. High Strain-Rate Tensile Testing and Viscoplastic Parameter Identification Using Microscopic High-Speed Photography. *Int. J. Plast.* **2004**, *20*, 561–575. [CrossRef]
31. Hung, P.C.; Voloshin, A.S. In-Plane Strain Measurement by Digital Image Correlation. *J. Braz. Soc. Mech. Sci. Eng.* **2003**, *25*, 215–221. [CrossRef]
32. Schwab, R.; Harter, A. Extracting True Stresses and Strains from Nominal Stresses and Strains in Tensile Testing. *Strain* **2021**, *57*, e12396. [CrossRef]

photonics

MDPI

Communication

Design of a Broadband Fiber Optic Mode Coupler for Multimode Optical Coherence Tomography

Dora Juan Juan Hu [1,2,*], Linbo Liu [2], Hui Dong [1] and Hailiang Zhang [1]

1 Institute for Infocomm Research, Agency for Science, Technology and Research, Singapore 138632, Singapore
2 School of Electrical and Electronic Engineering, Nanyang Technological University, Singapore 639798, Singapore
* Correspondence: jjhu@i2r.a-star.edu.sg

Abstract: In this paper, we propose an optical fiber-based broadband mode coupler for multimode optical coherence tomography (OCT) in the O-band (1.26–1.36 μm). The proposed device uses a tapered few-mode fiber (FMF) to lower the effective mode index of the selected higher-order mode, which can be phase matched to the fundamental mode of the single-mode fiber (SMF). The tapered FMF and the SMF are side polished to reduce the core-to-core separation to achieve efficient mode coupling. Key design parameters such as the tapering ratio of the FMF, FMF core to SMF core separation, coupler length, and coupling ratio in the O-band are studied thoroughly. Higher-order modes of the FMF will be effectively coupled from the fundamental mode of SMF in the sample arm of the multimode OCT system. The reflected signals of the higher-order modes from the sample will be separated into several single-mode signals using the same fiber device before interfering with the reference light, which was not possible before. The proposed fiber device will be a key component to efficiently achieve multimode OCT operation with better signal collection efficiency and improved penetration depth for deep tissue imaging.

Keywords: multimode optical coherence tomography; few-mode fiber (FMF); broadband fiber optic mode coupler

1. Introduction

Optical coherence tomography (OCT) has been established as a standard-of-care tool for the diagnosis of a wide spectrum of ocular diseases and a powerful tool for the diagnosis of coronary artery disease and gastrointestinal diseases [1–3]. However, penetration depth and resolution degradation in deep tissue are two major limitations of the technology for existing and potential clinical use [4,5]. For example, OCT is very limited in imaging the choroid, sclera, and optic nerve head due to limited penetration depth, making it unfit for examining abnormalities in these tissues [6–9]. In intravascular and endoscopic applications, the penetration limit precludes the technology for assessing deep lesions such as invasive tumors in digestive tracts and large plaques in coronary arteries [10–13]. In the single mode regime, OCT has reached shot-noise limited detection, which is the fundamental limitation on penetration given a certain maximum permissible exposure [14–18]. The way to improve penetration depth is to shift to the multimode regime, where signal collection efficiency can be improved by an order of magnitude by collecting backscattered light from tissue with a larger fiber core size and larger numerical aperture. Few mode fiber (FMF)-based simultaneous bright and dark field OCT imaging has demonstrated good prospects to achieve improved detection efficiency, although still suffering from crosstalk and sensitivity drop due to mode mismatch between the higher-order modes and the reference [19,20]. The key to addressing these limitations in multimode OCT is to separate OCT signals of different modes with a mode-selective device and detect them individually. More recently, a modally specific photonic lantern was used to achieve mode-dependent coupling and solved the sensitivity issue in multimode OCT at 930 nm [21]. However, 930 nm is not a commonly used center wavelength where the quantum efficiency of either Si or InGaAs-based linear cameras is low. In this paper, we present the design of a detection module with mode selectivity for the multimode OCT in 1.26–1.36 μm, typically used for OCT imaging applications. In the simulation study, a two-mode FMF is used to construct the mode-selective device. The simulation results suggest that the proposed few-mode detection module could enable multimode OCT detection, totaling approximately 1.69 times of conventional OCT signal.

2. Operating Principle of the Proposed Multimode OCT System and the Broadband Fiber Optic Mode Coupler

In the standard OCT operating in a single-mode regime, only backscattered (ballistic) photons coupled into the LP_{01} mode are detected. Different from the standard OCT, in the proposed multimode OCT setup, an FMF guides both LP_{01} and higher-order modes to the mode-selective coupler, which separates the signals of multiple modes with high coupling efficiency. For simplicity, we chose a two-mode FMF in the study, which only supports LP_{01} and LP_{11} modes.

In the proposed multimode OCT setup, a portion of the back-reflected or backscattered light from the sample will be picked up by the beam splitter (BS) and coupled into FMF. This portion of the signal is coupled to the FMF in both LP_{01} and LP_{11} modes, which are shown as the red rays for LP_{01} mode and black rays for LP_{11} mode reflected by the BS in Figure 1a. Compared with the standard single-mode detection, there is a significant portion of the signal in the LP_{11} mode carrying useful information that can be detected in the proposed multimode OCT system. The rest of the sample light, i.e., the backscattered light, is directed to pass the BS and coupled back to the same SMF that delivers the excitation light, as shown in the red rays in Figure 1a.

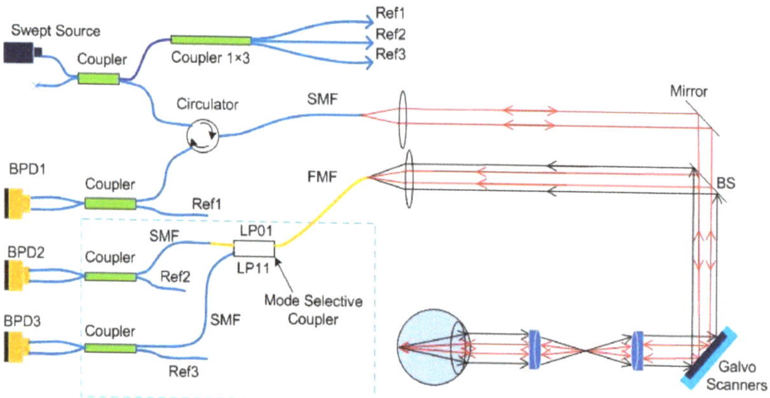

(a)

Figure 1. *Cont.*

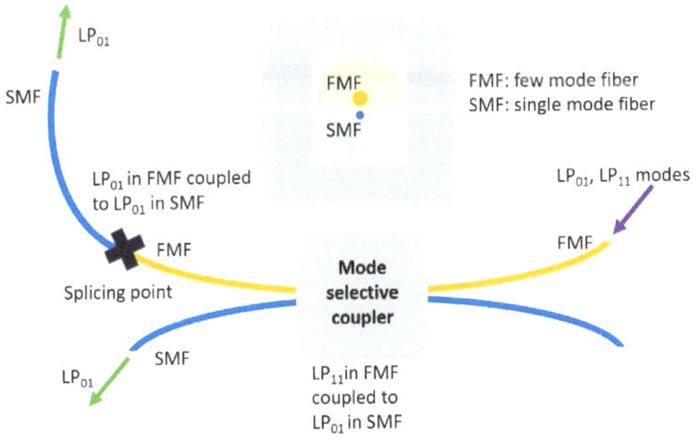

(b)

Figure 1. (a) Multimode OCT schematic with the FMF-based mode-selective coupler. SMF: single mode fiber; FMF: few-mode fiber; BPD: balanced photodetectors; BS: beam splitter; (b) Proposed FMF detection module.

The detection of signals of both LP_{01} and LP_{11} modes will result in better signal collection efficiency compared with the standard single-mode OCT, where only the signal of LP_{01} mode is detected. Since the OCT noise is dominated by the shot noise of the reference light, i.e., short noise limited, the sensitivity (signal-to-noise ratio with a perfect reflector) is solely dependent on the signal detected [18]. On the other hand, the penetration depth of OCT is proportional to the sensitivity.

The reference light is split into three beams, and each of the reference beams will independently interfere with the signals guided in the excitation SMF, the LP_{01} mode of FMF, and the LP_{11} mode of FMF, respectively. The three interference signals are separately detected by the use of three balanced detectors.

FMF has been used extensively in communication applications to augment the transmission capacity and for selective mode couplers [22–26]. For example, an FMF-based mode-selective coupler was developed to achieve mode division multiplexing from 1515–1590 nm. The fundamental mode of the SMF LP_{01} mode is selectively coupled to higher-order modes in FMF for transmission [26]. There are two parts of detection from mode coupling in the proposed FMF detection module, as shown in Figure 1b. Firstly, a tapered FMF is used to phase match its LP_{11} mode to the LP_{01} mode in SMF. In this study, the FMF only supports two modes for simplicity. If FMF supports more than one higher-order mode, each higher-order mode needs to be selectively phase matched to the LP_{01} mode in SMF for selective mode coupling. Both FMF and SMF are side polished to reduce the core-to-core separation to achieve effective mode coupling between "the LP11 in FMF coupled to LP01 in SMF" occurring in the mode-selective coupler; the corresponding output port (lower left SMF in Figure 1b) is guiding LP_{01} mode in SMF to interfere with reference. Secondly, at the other output port of the coupler module (upper left FMF in Figure 1b), the LP_{01} mode in FMF is coupled to LP_{01} mode in SMF through butt coupling or fusion splicing [27]. Then the LP_{01} mode in SMF will interfere with another reference. For desirable OCT imaging applications, the detection module is targeted to achieve broadband selective mode coupling in the O-band (1.26–1.36 μm).

3. Results and Discussion

In this section, the phase matching conditions and the selective mode coupling between the FMF and SMF in the proposed selective mode coupler structure is presented. Firstly, the side coupling between the tapered FMF LP_{11} mode and SMF LP_{01} mode is modeled by two parallel waveguides of FMF and SMF with the simulation parameters as shown in Table 1, including the core diameters, the refractive index of the cladding, which is calculated based on Sellmeier equation for fused silica to account for material dispersion, numerical aperture (NA) which can be used to calculate the refractive index of the core, the tapering ratio of the FMF, SMF core to FMF core separation, wavelength band, and coupler length. The Sellmeier equation used in this work is $n_{clad}(\lambda) = \sqrt{1 + \frac{0.6961633\lambda^2}{\lambda^2 - 0.0684043^2} + \frac{0.4079426\lambda^2}{\lambda^2 - 0.1162414^2} + \frac{0.8974794\lambda^2}{\lambda^2 - 9.896161^2}}$ [28], where λ is the wavelength with the unit of μm, and n_{clad} is the refractive index of the cladding. The refractive index of the core is calculated by $n_{core} = \sqrt{n_{clad}^2 + NA^2}$. In this study, a commercial software (BeamProp®—RSOFT Design®) was used to model the beam propagation and mode coupling in the optical coupler device.

Table 1. Simulation parameters in the side coupling.

Fiber Parameters	SMF	FMF
Core diameter (untapered)	8 μm	19 μm
Refractive index of the cladding	Sellmeier Equation for silica	Sellmeier Equation for silica
Numerical aperture (NA)	0.12	0.12
Tapering ratio	1	0.7–0.9
SMF core to FMF core separation	4–4.9 μm	
Wavelength	1.26–1.36 μm	
Coupler length	Up to 20 mm	

The effective mode indices in FMF and SMF are shown in Figure 2. The mode curves in untapered FMF are depicted by red and yellow curves, which are higher than the LP_{01} mode index curve of the SMF. Tapered FMF has lower mode indices compared with untapered FMF. The phase match condition is satisfied when the LP_{01} mode curve in SMF and the LP_{11} mode curve in tapered FMF are intersected in the O-band, as depicted by the blue curve and green circles curve, respectively. The tapered FMF has a core diameter of 15.1 μm

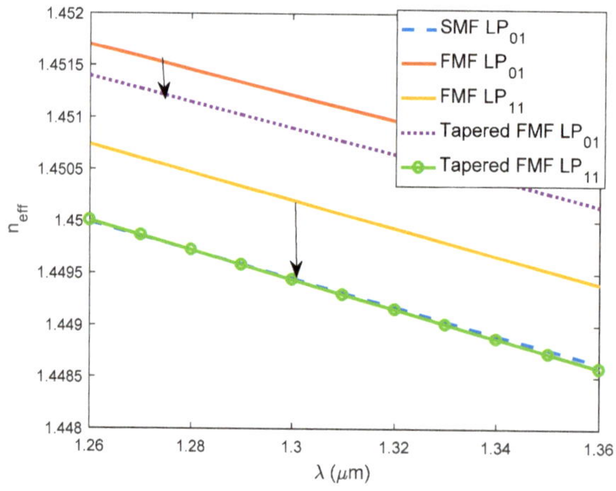

Figure 2. Phase matching curves of the modes in FMF and SMF.

The mode coupling between the two parallel coupled waveguides can be analyzed by coupled mode theory [29]. The SMF core to FMF core distance is a critical parameter for the mode coupling in two parallel coupled waveguides as the coupling coefficient

k_c is inversely dependent on it [30,31]. In addition, the coupling length is dependent on the coupling coefficient. Thus, it is also dependent on the core-to-core separation. The coupling ratio η in this dissimilar fiber coupler determines how much power is transferred from one core to the other, so it is essential to analyze the wavelength-dependent coupling ratio to ensure sufficient coupling for all wavelengths in the chosen band. The coupling between the LP_{11} mode in tapered FMF and the LP_{01} mode in SMF is analyzed by the beam propagation method. The normalized launch field is LP_{11} mode in the FMF core, and the mode power in both the FMF core and the SMF core is monitored along the propagation length. As shown in Figure 3a, the normalized mode power of LP_{11} mode in the FMF core (solid curves) is coupled to LP_{01} mode in SMF (dashed and dotted curves) and vice versa along the entire length of the coupler. The core-to-core separation is 4 μm in this simulated structure. The maximum coupling occurs at the coupling length L_c, corresponding to the maximum achievable coupling ratio η or the normalized coupled mode power of LP_{01} mode in SMF. The mode coupling curves are plotted at two wavelengths of 1.26 and 1.36 μm, with different L_c and η values of 5575 μm, 0.94 and 8210 μm, 0.97, respectively. Figure 3b shows the mode coupling curves at wavelength of 1.26 μm for coupler structures with core-to-core separation d of 4 μm and 4.9 μm. Clearly, when d is larger, both cores are further apart, leading to a weaker coupling thus longer coupling length L_c of 8210 μm associated with η of 0.88.

The coupling length L_c, and the maximum coupling ratio η, present dependence on the core-to-core separation d and vary in different wavelengths as shown in Figure 4a,b, respectively. In general, coupler structures with larger d values have weaker coupling coefficients, thus, are associated with longer coupling length L_c for maximum power transfer [29]. As shown in Figure 4b, all the maximum power transfers occur at 1.31 μm, as the phase matching condition is satisfied at this wavelength.

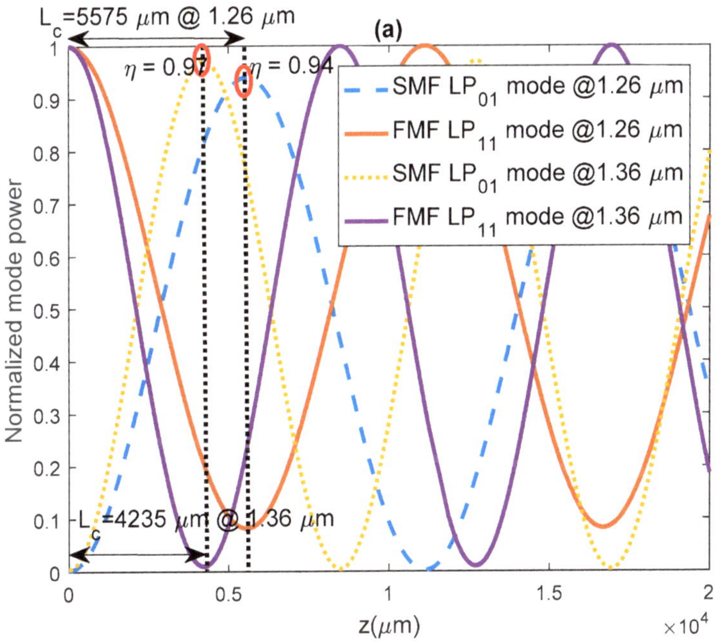

Figure 3. *Cont.*

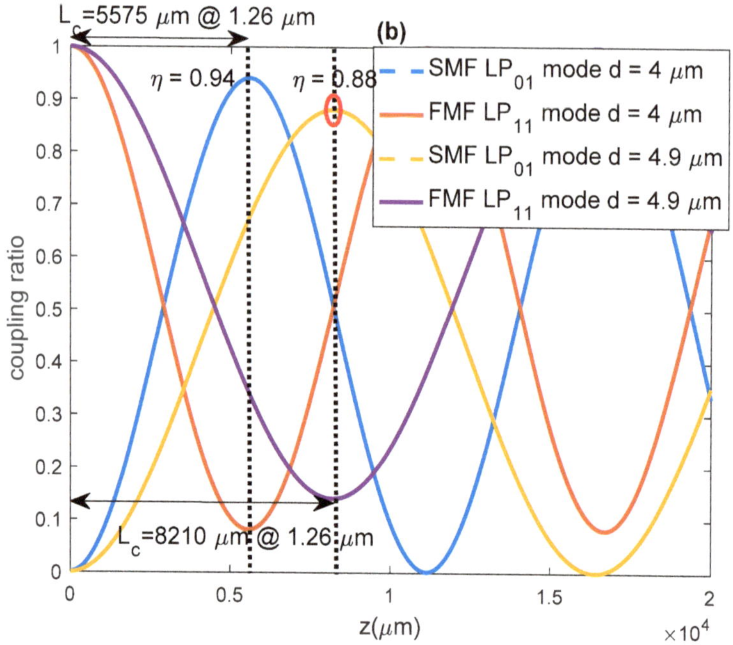

Figure 3. (**a**): Mode coupling varies with coupler length wavelength; (**b**) Mode coupling varies with core-to-core separation.

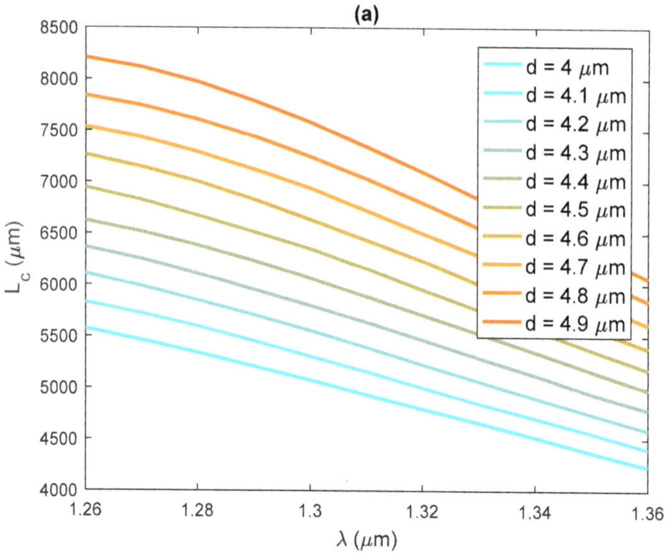

Figure 4. *Cont.*

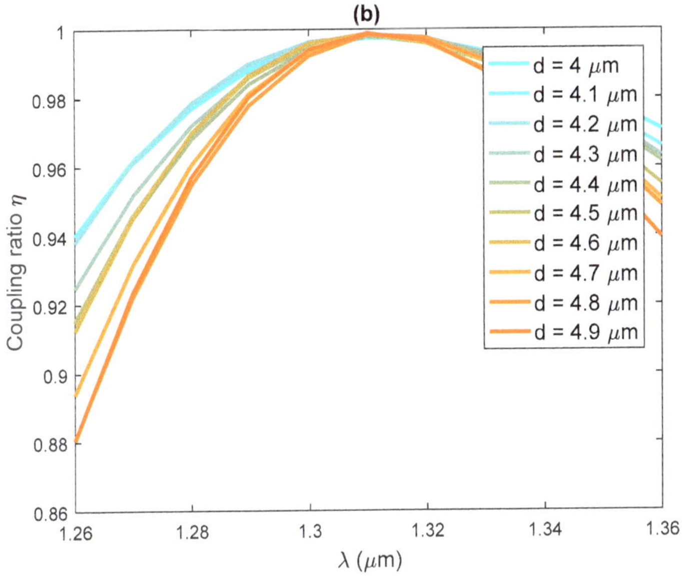

Figure 4. (**a**): Coupling length as a function of wavelength and core-to-core separation d; (**b**) coupling ratio η as a function of wavelength and core-to-core separation d.

The coupler length L is another critical design parameter shown in the colormap of the broadband coupling ratio η in Figure 5a. The simulated coupler structure has core-to-core separation d = 4 μm. The two solid lines mark the permissible coupler length range to ensure the chosen coupler structure achieves a broadband coupling ratio above a threshold value of 0.9. For other selected threshold values of the broadband coupling ratio varying from 0.9 to 0.55 with an interval of 0.05 and varying core-to-core separation from 4 to 4.9 μm, the allowed coupler length range is plotted in Figure 5b. As the threshold value increases, the range of the coupler length becomes narrower. For a high coupling ratio threshold value of 0.9, the selection of core-to-core diameter values is critical as the maximum and minimum L curves close at d around d = 4.2 μm. In other words, coupler structures with d values above 4.2 μm would not provide a broadband coupling ratio at 0.9 regardless of the coupler length.

For a fixed coupler length, e.g., L = 5 mm, the coupling ratio in the O-band as a function of the core-to-core separation d varying from 4 to 4.9 μm is plotted in Figure 6a. Couplers with smaller separations are associated with a higher coupling ratio and are more desirable to achieve broadband coupling above a chosen threshold value. Specifically, the coupler with d = 4 μm can achieve broadband coupling above 0.9, as shown by the solid blue curve. For estimation purposes, assuming the LP_{11} mode power is equivalent to LP_{01} mode power and assuming the BS ratio is 90:10, the multimode detection using the selective mode coupler with output to BPD2 in the multimode OCT system as shown in Figure 1a could achieve 0.81 ($0.9 \times 0.9 = 0.81$) or more for all wavelengths in the O-band.

The coupling of FMF LP_{01} to SMF LP_{01} mode is through direct fusion splicing between both fibers. The coupling ratio varies from 0.87 to 0.89 for tapered FMF with a diameter of 15.1 μm, whereas it varies from 0.74 to 0.77 for untapered FMF, as shown in Figure 6b. The LP_{01} mode detection using a conventional OCT signal is normalized to 1. The LP_{01} mode detection through two paths, by BPD1 and BPD3, could achieve greater than 0.88 ($0.1 + 0.9 \times 0.87 \approx 0.88$) if using tapered FMF butt coupled to SMF. The total detected power in the multimode OCT system is approximately 1.69 times of the conventional OCT signal.

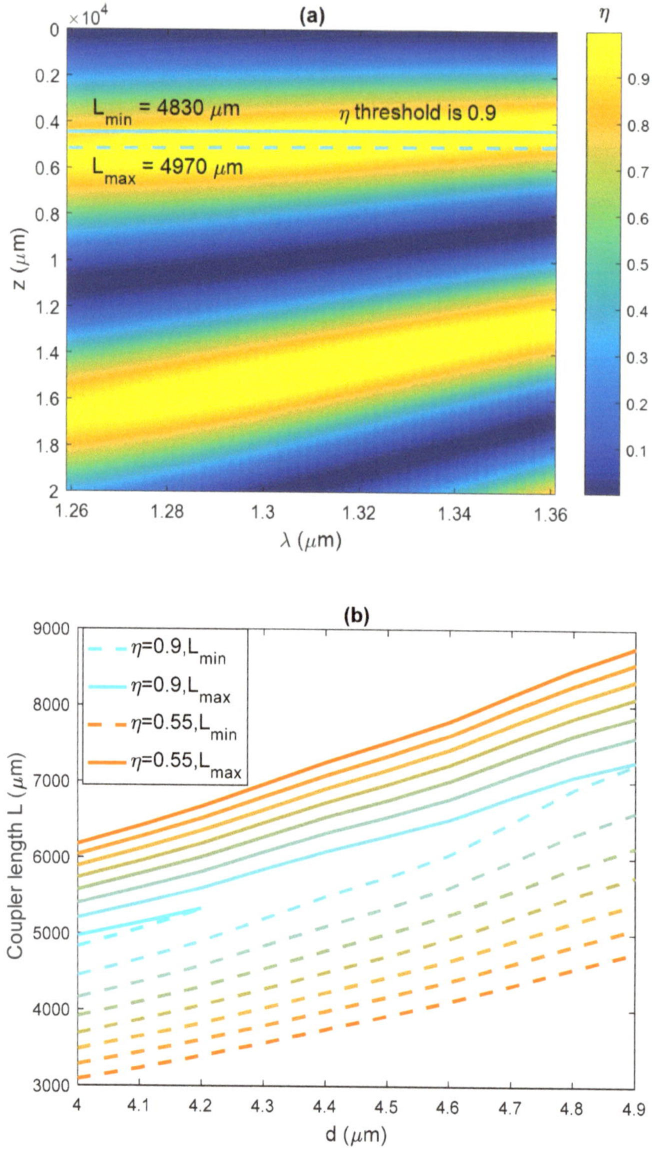

Figure 5. (**a**): Color map of the coupling ratio η as a function of coupler length and wavelength. Two lines represent the allowed coupler length range to achieve the broadband coupling ratio 0.9; (**b**) Allowed coupler length range for chosen coupling ratio threshold, varying from 0.9 to 0.55 with an interval of 0.05.

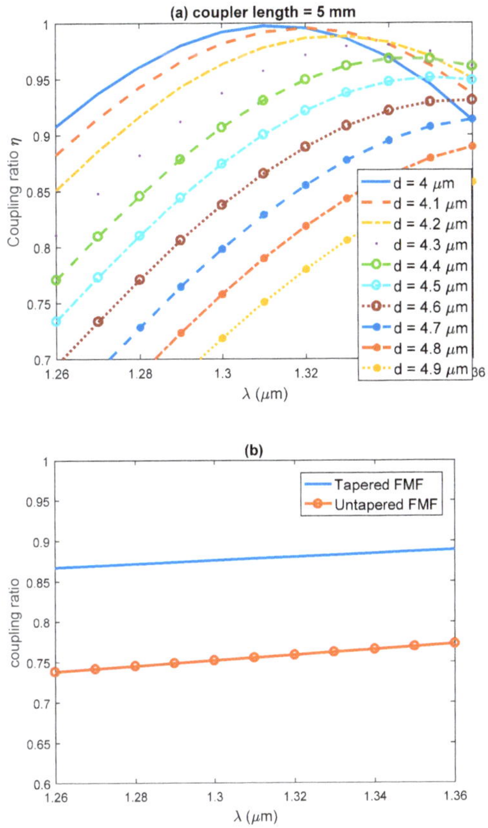

Figure 6. **(a)** Broadband coupling of the device with varying core-to-core separation d, from 4 to 4.9 μm, and a fixed coupler length L of 5 mm. (**b**) Mode coupling of SMF LP_{01} to FMF LP_{01} mode in tapered FMF (solid line) and untapered FMF (solid line with circles) through direct fusion splicing.

The FMF LP_{01} mode and SMF LP_{01} mode are not phase-matched. Thus, the coupling efficiency would be very small. For example, the experiment result reported in a previous paper shows the coupling efficiency was 0.1% between the FMF LP_{01} mode and SMF LP_{01} mode as compared to the coupling efficiency of 78.9% for desirable mode coupling between the FMF LP_{11} mode and SMF LP_{01} mode at 1550 nm [26]. In addition, the noise primarily comes from the reference light. The mode coupling is restricted to the sample arm, and there is little effect on the imaging for two reasons. First of all, the optical path length difference between the crosstalk and the reference is very large so that the interference fringes due to crosstalk will not be detected by the detection electronics since it is essentially a low-pass filter. Second, for biomedical applications, in most cases, the sample reflectivity is a few orders of magnitude smaller than that of the reference light. Therefore, the sample power (or its variations) can be negligible in calculating the noise. According to Eq. 2 in Nassif et al. [18], the overall noise in electrons squared per readout cycle and per detector element is given by:

$$\sigma^2 = \sigma_{r+d}^2 + \frac{\eta e^2 P_{ref} \tau_i}{E_v} + \left(\frac{\eta e^2 P_{ref} \tau_i}{E_v} \right)^2 \frac{1}{\tau_i} \tau_{coh} \left[e^2 \right]$$

where e is the electron charge; P_{ref} is the reference arm power per detector element at the detection arm fiber tip; τ_i is the integration time, and E_v is the photon energy. σ^2_{r+d} is the electrical noise term, $\frac{\eta e^2 P_{ref}\tau_i}{E_v}$ is the shot noise term, which is equal to the electrons from the reference light, and the rest of the right-hand side is the RIN term. In shot-noise OCT devices, the equation above can be reduced to $\frac{\eta e^2 P_{ref}\tau_i}{E_v}$.

Therefore, the improvement in SNR is the improvement in signal power, which is dependent on the beam-splitting ratio. The light attenuation in turbid media in biological tissue is normally modeled with the Beer-Lambert law, which means the penetration depth is approximately linearly related to the SNR. However, the improvement in penetration depth is dependent on the attenuation coefficient of the sample, so without knowledge of the sample, it is not possible to quantify penetration improvement. However, a simpler way to understand the advantage of the proposed method is in terms of imaging speed. For example, with a 90:10 beam splitter, the signal power will be 69% higher, thus improving SNR by 69%.

If FMF supports more than one higher-order mode, e.g., in addition to LP_{11} mode, LP_{21} and LP_{02} or even higher-order modes are also supported, the proposed FMF detection module will be modified to cascade additional mode-selective coupler to selectively phase match each higher-order mode to the LP_{01} mode in SMF for selective mode coupling in the O-band. In each mode-selective coupler, the tapering ratio of the FMF and the core-to-core separation need to be optimized specifically for the coupling between the selective higher-order mode of FMF and the LP_{01} mode in SMF. The optimization for other higher-order modes can be performed by the same procedure presented in this work. Taking the assumption of equal power distribution among all the optical modes in the FMF, the FMF detection module can be designed to achieve broadband mode coupling of coupling efficiency over η between each higher-order mode and the LP_{01} mode in SMF, e.g., for an N higher-order mode FMF, the proposed detection module could achieve $0.88 + N\eta \times 0.9$ times of conventional OCT signal. However, it should be noted that propagating more higher-order modes in the FMF demands a good extinction ratio in the mode-selective coupler. The extinction ratio determines the ratio of the selective mode coupling to the desired mode compared to unwanted modes [26]. The extinction ratio will be a critical parameter to minimize crosstalk among higher-order modes at the detector, and thus it needs further study in the FMF coupler design optimization.

The presented results are based on simulations of the optimized FMF coupler only. We have not built the system yet. We will report the experimental results in future publications.

4. Conclusions

Our preliminary simulation results demonstrate that multimode detection in OCT can be efficiently realized using a broadband all-fiber mode coupler in the O-band. It enhances signal strength as the simulation demonstrates over 1.69 times of conventional OCT signal could be achieved. This approach will also improve penetration depth in OCT imaging. Furthermore, the proposed detection module makes it possible to suppress speckle noise by incoherent averaging OCT images of different modes as their speckles are not correlated.

Author Contributions: Conceptualization, D.J.J.H., and L.L.; Investigation and writing—original draft preparation, D.J.J.H.; writing—review and editing, D.J.J.H., L.L., H.D. and H.Z. All authors have read and agreed to the published version of the manuscript.

Funding: We acknowledge the funding support from the Singapore Ministry of Health's National Medical Research Council under its Open Fund Individual Research Grant (MOH-000384) and the Ministry of Education Singapore under its Academic Research Funding Tier 2 (MOE-T2EP30120-0001) and Academic Research Funding Tier 1 (RG35/22).

Institutional Review Board Statement: Not applicable.

Informed Consent Statement: Not applicable.

Data Availability Statement: Not applicable.

Conflicts of Interest: The authors declare no conflict of interest.

References

1. Huang, D.; Swanson, E.A.; Lin, C.P.; Schuman, J.S.; Stinson, W.G.; Chang, W.; Hee, M.R.; Flotte, T.; Gregory, K.; Puliafito, C.A. Optical coherence tomography. *science* **1991**, *254*, 1178–1181. [CrossRef] [PubMed]
2. Drexler, W.; Liu, M.; Kumar, A.; Kamali, T.; Unterhuber, A.; Leitgeb, R.A. Optical coherence tomography today: Speed, contrast, and multimodality. *J. Biomed. Opt.* **2014**, *19*, 071412. [CrossRef]
3. Vakoc, B.J.; Lanning, R.M.; Tyrrell, J.A.; Padera, T.P.; Bartlett, L.A.; Stylianopoulos, T.; Munn, L.L.; Tearney, G.J.; Fukumura, D.; Jain, R.K.; et al. Three-dimensional microscopy of the tumor microenvironment in vivo using optical frequency domain imaging. *Nat. Med.* **2009**, *15*, 1219–1223. [CrossRef] [PubMed]
4. Cogliati, A.; Canavesi, C.; Hayes, A.; Tankam, P.; Duma, V.-F.; Santhanam, A.; Thompson, K.P.; Rolland, J.P. MEMS-based handheld scanning probe with pre-shaped input signals for distortion-free images in Gabor-domain optical coherence microscopy. *Opt. Express* **2016**, *24*, 13365–13374. [CrossRef] [PubMed]
5. Wang, R.K. Signal degradation by multiple scattering in optical coherence tomography of dense tissue: A Monte Carlo study towards optical clearing of biotissues. *Phys. Med. Biol.* **2002**, *47*, 2281–2299. [CrossRef]
6. Park, H.-Y.L.; Lee, N.Y.; Choi, J.A.; Park, C.K. Measurement of scleral thickness using swept-source optical coherence tomography in patients with open-angle glaucoma and myopia. *Am. J. Ophthalmol.* **2014**, *157*, 876–885. [CrossRef] [PubMed]
7. Ikuno, Y.; Nouchi, T.; Yasuno, Y. Scleral and Choroidal Thickness in Normal Subjects Measured by High-Penetration Optical Coherence Tomography With Long Wavelength. *Investig. Ophthalmol. Vis. Sci.* **2010**, *51*, 1634.
8. Willemse, J.; Gräfe, M.G.O.; Verbraak, F.D.; Boer, J.F.d. In Vivo 3D Determination of Peripapillary Scleral and Retinal Layer Architecture Using Polarization-Sensitive Optical Coherence Tomography. *Transl. Vis. Sci. Technol.* **2020**, *9*, 21. [CrossRef]
9. Spaide, R.F.; Fujimoto, J.G.; Waheed, N.K.; Sadda, S.R.; Staurenghi, G. Optical coherence tomography angiography. *Prog. Retin. Eye Res.* **2018**, *64*, 1–55. [CrossRef]
10. Yoshida, S.; Tanaka, S.; Hirata, M.; Mouri, R.; Kaneko, I.; Oka, S.; Yoshihara, M.; Chayama, K. Optical biopsy of GI lesions by reflectance-type laser-scanning confocal microscopy. *Gastrointest. Endosc.* **2007**, *66*, 144–149. [CrossRef]
11. Bezerra, H.G.; Costa, M.A.; Guagliumi, G.; Rollins, A.M.; Simon, D.I. Intracoronary Optical Coherence Tomography: A Comprehensive Review. *JACC Cardiovasc. Interv.* **2009**, *2*, 1035–1046. [CrossRef] [PubMed]
12. Liu, L.; Gardecki, J.A.; Nadkarni, S.K.; Toussaint, J.D.; Yagi, Y.; Bouma, B.E.; Tearney, G.J. Imaging the subcellular structure of human coronary atherosclerosis using micro–optical coherence tomography. *Nat. Med.* **2011**, *17*, 1010–1014. [CrossRef] [PubMed]
13. Lee, H.C.; Ahsen, O.O.; Liang, K.; Wang, Z.; Figueiredo, M.; Giacomelli, M.G.; Potsaid, B.; Huang, Q.; Mashimo, H.; Fujimoto, J.G. Endoscopic optical coherence tomography angiography microvascular features associated with dysplasia in Barrett's esophagus (with video). *Gastrointest. Endosc.* **2017**, *86*, 476–484.e3. [CrossRef]
14. White, B.R.; Pierce, M.C.; Nassif, N.; Cense, B.; Park, B.H.; Tearney, G.J.; Bouma, B.E.; Chen, T.C.; de Boer, J.F. In vivo dynamic human retinal blood flow imaging using ultra-high-speed spectral domain optical Doppler tomography. *Opt. Express* **2003**, *11*, 3490–3497. [CrossRef]
15. Leitgeb, R.; Hitzenberger, C.K.; Fercher, A.F. Performance of fourier domain vs. time domain optical coherence tomography. *Opt. Express* **2003**, *11*, 889–894. [CrossRef] [PubMed]
16. You, J.-W.; Chen, T.C.; Mujat, M.; Hyle Park, B.; de Boer, J.F. Pulsed illumination spectral-domain optical coherence tomography for human retinal imaging. *Opt. Express* **2006**, *14*, 6739–6748. [CrossRef] [PubMed]
17. Yun, S.H.; Tearney, G.J.; Bouma, B.E.; Park, B.H.; de Boer, J.F. High-speed spectral-domain optical coherence tomography at 1.3 μm wavelength. *Opt. Express* **2003**, *11*, 3598–3604. [CrossRef]
18. Nassif, N.; Cense, B.; Hyle Park, B.; Yun, S.H.; Chen, T.C.; Bouma, B.E.; Tearney, G.J.; de Boer, J.F. In vivo human retinal imaging by ultrahigh-speed spectral domain optical coherence tomography. *Opt. Lett.* **2004**, *29*, 480–482. [CrossRef]
19. Eugui, P.; Lichtenegger, A.; Augustin, M.; Harper, D.J.; Fialová, S.; Wartak, A.; Hitzenberger, C.K.; Baumann, B. Few-Mode Fiber Detection for Tissue Characterization in Optical Coherence Tomography. In Proceedings of the Optical Coherence Imaging Techniques and Imaging in Scattering Media II, Munich Germany, 25–29 June 2017; Wojtkowski, M., Ed.;
20. Eugui, P.; Lichtenegger, A.; Augustin, M.; Harper, D.J.; Muck, M.; Roetzer, T.; Wartak, A.; Konegger, T.; Widhalm, G.; Hitzenberger, C.K.; et al. Beyond backscattering: Optical neuroimaging by BRAD. *Biomed. Opt. Express* **2018**, *9*, 2476–2494. [CrossRef]
21. Sivry-Houle, M.P.d.; Beaudoin, S.B.; Brais-Brunet, S.; Dehaes, M.; Godbout, N.; Boudoux, C. All-fiber few-mode optical coherence tomography using a modally-specific photonic lantern. *Biomed. Opt. Express* **2021**, *12*, 5704–5719. [CrossRef]
22. Chang, S.H.; Moon, S.-R.; Chen, H.; Ryf, R.; Fontaine, N.K.; Park, K.J.; Kim, K.; Lee, J.K. All-fiber 6-mode multiplexers based on fiber mode selective couplers. *Opt. Express* **2017**, *25*, 5734–5741. [CrossRef]
23. Corral, J.L.; Garcia-Rodriguez, D.; Llorente, R. Mode-Selective Couplers for Two-Mode Transmission at 850 nm in Standard SMF. *IEEE Photonics Technol. Lett.* **2016**, *28*, 425–428. [CrossRef]
24. Li, A.; Chen, X.; Amin, A.A.; Shieh, W. Fused Fiber Mode Couplers for Few-Mode Transmission. *IEEE Photonics Technol. Lett.* **2012**, *24*, 1953–1956.

25. Bremer, K.; Schlangen, S.; Böhm, S.; Wellmann, F.; Steinke, M.; Neumann, J.; Roth, B.; Overmeyer, L. Investigation of grating assisted mode-selective few-mode fused fiber couplers. In Proceedings of the 2019 Conference on Lasers and Electro-Optics Europe and European Quantum Electronics Conference, Munich Germany, 23–27 June 2019.
26. Park, K.J.; Song, K.Y.; Kim, Y.K.; Lee, J.H.; Kim, B.Y. Broadband mode division multiplexer using all-fiber mode selective couplers. *Opt. Express* **2016**, *24*, 3543–3549. [CrossRef] [PubMed]
27. Hu, D.J.J.; Liu, L.; Dong, H.; Zhang, H. Design of a Tapered Few-Mode Fiber based Broadband Mode Coupler for Multimode OCT. In Proceedings of the Imaging and Applied Optics Congress 2022 (3D, AOA, COSI, ISA, pcAOP), Vancouver, BC, Canada, 11–15 July 2022.
28. Malitson, H. Interspecimen Comparison of the Refractive Index of Fused Silica*,†. *J. Opt. Soc. Am.* **1965**, *55*, 1205–1209. [CrossRef]
29. Okamoto, K. *Fundamentals of Optical Waveguides*; Academic press: Cambridge, MA, USA, 2010.
30. Hu, D.J.J.; Shum, P.P.; Lim, J.L.; Cui, Y.; Milenko, K.; Wang, Y.; Wolinski, T. A Compact and Temperature-Sensitive Directional Coupler Based on Photonic Crystal Fiber Filled With Liquid Crystal 6CHBT. *IEEE Photonics J.* **2012**, *4*, 2010–2016. [CrossRef]
31. Xu, Z.; Lim, J.; Hu, D.J.J.; Sun, Q.; Wong, R.Y.-N.; Li, K.; Jiang, M.; Shum, P.P. Investigation of temperature sensing characteristics in selectively infiltrated photonic crystal fiber. *Opt. Express* **2016**, *24*, 1699–1707. [CrossRef]

photonics

MDPI

Article

Ensquared Energy and Optical Centroid Efficiency in Optical Sensors: Part 1, Theory

Marija Strojnik [1,*], Beethoven Bravo-Medina [2], Robert Martin [3,†] and Yaujen Wang [4,‡]

1 Optical Research Center, Apdo. Postal 1-948, Leon 37000, Mexico
2 Centro Universitario de Ciencias Exactas e Ingenierías, University of Guadalajara, Guadalajara 44840, Mexico
3 Independent Researcher, West Hills, CA 91304, USA
4 Independent Researcher, Arcadia, CA 91006, USA
* Correspondence: mstrojnik@gmail.com; Tel.: +1-480-479-7817
† When this work was performed, the author's affiliation was Lockheed-Martin Missiles & Space Company, Sunnyvale, CA 94088, USA.
‡ When this work was performed, the author was affiliated with Northrop Grumman, Aerospace Systems, Azusa, CA 91702, USA.

Abstract: High-performance megapixel focal plane arrays with small pixels have been widely used in modern optical remote sensing, astronomical, and surveillance instruments. In the prediction models applied in the traditional instrument performance analysis, the image of a point source is assumed to fall on the center of a detector pixel. A geometrical image of a point source in the realistic optical system may actually fall on any position on the detector pixel because the sensor's line of sight includes pointing errors and jitter. This traditional assumption may lead to an optimistic error, estimated at between 10% and 20%. We present the critical factors that impact the performance estimate in a realistic instrument design based on the prediction for the noise-equivalent power (*NEP*). They are the optical centroid efficiency (*OCE*) and the ensquared energy, or more precisely, the energy on the rectangular detector pixel (*EOD*). We performed the simulation studies for imaging with an optical system with and without a generalized rectangular central obscuration.

Keywords: optical system performance; image quality; figure of merit; radiometry; optical centroid efficiency, *OCE*; ensquared energy; energy on detector, *OED*; sensors; optomechatronics; optical devices

1. Discrete Pixels

The advances in IR sensor technology are arising to a large degree due to the improvements in focal plane array (*FPA*) technologies. These include an increase in FPA size, reduced detector pixel dimensions, low dark current, low readout noise, and high responsivity. A larger *FPA* covers a wider footprint in the object space, while a smaller pixel may lead to a reduction in the object-space footprint. For example, the Teledyne H2RG FPA, integrated into the James Webb Space Telescope (JWST), features a 2048 by 2048-pixel array with an 18 μm pixel pitch (distance between pixel centers) [1]. Raytheon produced a 2048 × 2048 detector array with a 20 μm-pixel pitch for the UK Visible and Infrared Survey Telescope for Astronomy (VISTA) [2]. The analysis presented here applies to the instruments operating from the UV to the microwaves. The sensor figures of merit were originally developed consistently using radiometric concepts for the IR part of the electromagnetic spectrum, so we rely and build our theory on those.

Modern instrument design procedures continue to rely on parameters such as enclosed energy and the noise-equivalent power (*NEP*) for their performance assessment and prediction. In Figure 1, we illustrate how the previously important quantity encircled energy increases as a function of the radius of the enclosing circle. We can also read the approximate amount of energy that would be incident, or enclosed, within a (circular) pixel equal to the radial distance *r*.

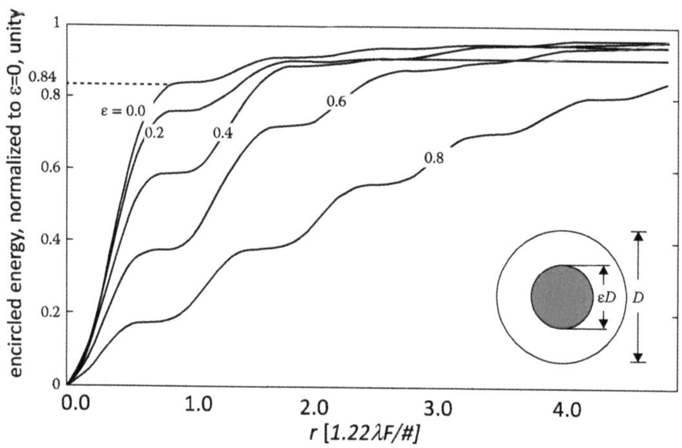

Figure 1. The normalized integrated, encircled energy as a function of radius r of the enclosing circle for an ideal, diffraction-limited optical system with a circular aperture. The obscuration ratio ε is a parameter (ratio of radii), varying from 0 (no obscuration) to 0.8 in increments of 0.2. The degree of compactness of the central spot may be recognized from the average slope of these curves for radial values less than 1. About 84% of the energy is enclosed within the first dark ring of a diffraction-limited optical system without the central obscuration. All the curves also flatten to zero slope for radial distances that correspond to the zero-rings, due to the absence of radiation there.

In Section 2, we introduce the concept of the energy on detector (*EOD*) to consider a rectangular pixelated image plane. We also account for the fact that the *EOD* is diminished when the pixel axis and optical system axis do not coincide, the real situation in most instruments. This leads us to introduce the optical centroiding efficiency (*OCE*). In Section 3, we develop a theoretical model to represent the imaging of a point source onto an arbitrary point on a detector pixel, to evaluate the *NEP* more realistically. We describe the relationship of the point spread function, *psf*, to both the *EOD* and the *OCE*. In the following section, the imaging theory is developed for the case when the optical system and the detector pixel are misaligned. In Section 4, the simulation methodology is described in detail. Next, we analyze the performance of an optical system with and without central obscuration. Finally in Section 6, we discuss the significance of using the OCE to predict the future performance of an instrument during its conceptual design stage. The conclusions are summarized in Section 7.

2. Energy Interception by Discrete Pixels

2.1. Energy on Detector (EOD)

The concept of the *EOD* became important when rectangular pixels were introduced, their dimensions became ever smaller, and their dimensions started to define the instrument resolution. Figure 2a presents the image that the optical system with a circular aperture delivers onto the focal plane, where an array of sensing elements detect the incident radiation. The inset shows the corresponding grayscale intensity, assuming low exposure. The first bright ring is barely visible. Theoretically, there is zero intensity in the first minimum, measured from the central peak.

This zero defines the first dark ring, or the radius of the resolution spot, according to the Rayleigh criterion, as illustrated in Figure 2b. In the 1980s, during the early periods of the design of the Mars orbiter, we required four pixels per resolution distance of the optical system to ensure the shape recognition of 1 m rocks or holes. Due to the 1 m diameter rover wheels, 1 m diameter holes were considered most detrimental, because the vehicle could become stuck inside. A hole would, then, bring its exploration to an end [3–5]. The orbiter was tasked to find a suitable lander landing site in the middle of an initial Mars

Rover exploration area. At that time, we were optimistically designing an autonomous rover with optical navigation system for autonomous navigation on its surface.

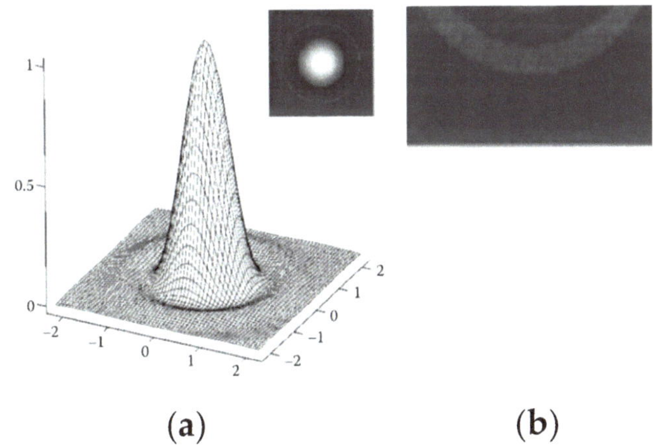

Figure 2. (**a**) Image by an optical system of a point object at infinity, with the central spot often referred to as the Airy disc; (**b**) Rayleigh resolution distance is equal to the Airy disc radius.

Figure 3 illustrates schematically how the pixel dimension relative to the size of the Airy disc affects the amount of energy that is incident on the detector pixel. The radius of the Airy disc is dependent on the optical system, $r_A = 1.22\ \lambda F/\#$. The wavelength is denoted by λ, and the instrument is characterized by its ratio $f/D = F/\#$. D is the aperture diameter, f is the focal distance, and F/# (sometimes also f/#) is the f-number. The pixel size is equal to the Airy disc diameter in the third illustration from the left. The two pixels on its left are larger by 20% and 40% than the diameter of the Airy disc. Both also include two point-sources positioned at the Rayleigh resolution limit. In this case, the pixel dimension acts to define the instrument resolution. The pixel on the furthest right would resolve the two point-sources at the Rayleigh's resolution limit, but only if it were correctly centered.

Figure 3. A schematic illustration of how the pixel dimension relative to the size of the Airy disc affects the amount of energy that is incident on and absorbed by the detector pixel.

We first modified the concept of energy enclosed within a circle, due to the existence of rectangular pixels. We then introduced a modified term due to the random position of the image centroid on the pixel. The *EOD* is a multiplicative correction factor that was invoked in the system design to account for the spread of an image of a point source over several detector pixels. The *EOD* may be formally defined as the fraction of energy incident on the signal-carrying detector pixel over the total energy within the image. This definition implicitly assumes that the centroid of the image of the point source is located at the pixel center. The pixel area is $4d_xd_y$ for the pixel dimensions of $2d_x$ by $2d_y$.

$$EOD(d_x, d_y) = \frac{\int_{-d_y}^{d_y} \int_{-d_x}^{d_x} psf(x,y)dx\,dy}{\int_{-\infty}^{\infty} \int_{-\infty}^{\infty} psf(x,y)dx\,dy} \quad [\text{unitless}] \tag{1}$$

Here, *psf(x,y)* is the point spread function of the optical system. In theory, it is unitless, but in practice, it is measured in units of [watts/cm^2] or [#photons/(cm^2 s)] [6–8]. We note that the (degree of) coherence of the object point source is not of interest. The source coherence may only change the form of the *psf(x,y)* function to some degree, but not the theoretical development presented here. For the image centered at the origin (0,0), the *psf(x,y)* attains the maximum value at $x = y = 0$ in this *EOD* formulation. The denominator is the total energy in the image of a point source. In practice, the total energy only includes the summation of energy incident on the few neighboring pixels.

In Figure 4, we see two images of point sources at infinity incident on the pixel distribution in the focal plane. In part (b), we see a pixel that is nearly centered on the optical axis, while in part (b), we show the displacement of the optical axis to the pixel corner. The image that is of approximately of the same size as the pixel (upper right, darker pinkish color) does not necessarily fall on a single pixel. The image centroid may be located anywhere on the specific pixel with equal probability. Even a significant fraction of its image, or energy, may be incident on the neighboring pixels.

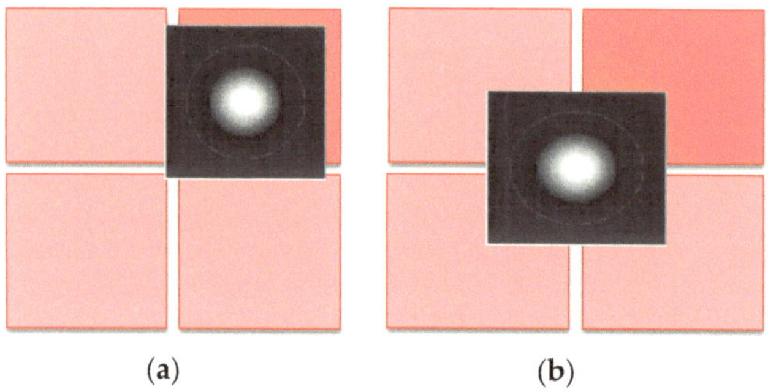

Figure 4. (**a**) The image of a point source is in general not aligned with a pixel center. (**b**) When in the extreme, limiting case, the image falls on the corner of four pixels, the detected signal in upper right pixel is equal to the noise in three neighboring pixels. In fact, there is no image detected against the noise, or possibly a broadly blurred image.

The pixel size is the most important parameter that determines the amount of the energy on detector, followed by the aperture configuration, and the aberrations that include any manufacturing and assembly errors. We consider a general aperture, incorporating an ideal circular aperture of diameter *D* with a rectangular obscuration, $0.27\,D \times 0.46\,D$, where *D* is the diameter of the entrance pupil. Figure 5a presents the aperture geometry, while part (b) shows the diffraction pattern in the image plane. The rectangular obscuration results in an asymmetrical *psf*. Such an obscuration has the effect of moving the energy out in an asymmetric fashion. The primary consequence of having an optical system with an obscured aperture is a decreased *EOD*.

The energy *EOD*, and, by derivation, the *NEP* and other figures of merit, were traditionally calculated under the assumption that the image of a point source on the optical axis, or the image centroid, is located exactly at the center of a detector pixel. This assumption would imply that the center of the central pixel of a large FPA is aligned within nanometers with the instrument optical axis. In the true and realistic operational environments, there exists too much image jitter for this assumption to be valid.

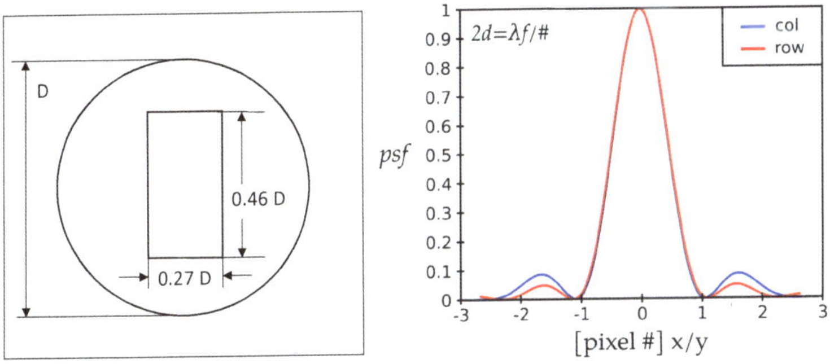

(a) (b)

Figure 5. (**a**) Transmission profile of a circular pupil featuring a general rectangular central obscuration. Gray indicates zero-transmission. The size of the obscuration rectangle is *0.27 D* × *0.46 D*, where *D* is the diameter of the entrance pupil. (**b**) Point spread function, *psf*, of the circular pupil with the internal central obscuration shown in part (**a**). Different profiles are obtained along the x- and y-dimensions because the rectangular obscuration is asymmetrical.

2.2. *Optical Centroiding Efficiency (OCE)*

The image of a point source is, in general, not aligned with a pixel center. In an extreme case, the image may fall on the corner of four pixels, as in Figure 4b. The energy in the signal-carrying upper-right pixel is equal to the noise in three neighboring pixels. In fact, there is no image detected against the noise.

The probability of finding the image centroid in the pixel center (that itself is located on the optical axis) is negligibly small. In the first approximation, we may reasonably assume that the position of the image centroid is uniformly distributed over the pixel area, defined as the product of the pitch along the two perpendicular directions. A correction analysis, hopefully leading to a single correction factor, is needed to formulate realistic design concepts applicable to the actual operational conditions. This is precisely what we endeavor to achieve here.

For a modern space/ground remote sensing instrument or a telescope with a large *FPA*, it is practically impossible to image a remote point source at the very center of a detector pixel, due to the image jitter and randomness in the line of sight. Toward a more realistic instrument modeling, we introduced a parameter, related to the *EOD*, that we call the optical centroiding efficiency (*OCE*). This statistical quantity tells us how much the *EOD* changes on average when the image centroid is displaced to a position on the pixel at $(\pm\Delta d_x, \pm\Delta d_y)$ where these two quantities lie between 0 and d_x, d_y, respectively.

3. Imaging Theory When Optical and Detector Axes Are Arbitrarily Displaced

3.1. *Image Radiometry of a Point Source*

The power collected from a distant point source by the aperture of the optical system in the image (the focal) plane, E_P, is most easily obtained from the power transfer equation [9,10]. This formulation considers a continuous detection process, such as the film used in the past, prior to the introduction of the digital sensors.

$$E_P = \Phi_P A_S \tau \quad [\mathrm{W}] \tag{2}$$

Here, Φ_P is the irradiance at the instrument aperture [watts/cm^2] or [#photons/(cm^2 s)]. A_s is the sensor/instrument aperture area [cm^2]; τ is the instrument optical transmission factor, including any aperture obscuration or obstruction.

When the image of a point object in the object space is spread over several neighboring detector pixels in the image space, only the central pixel contributes to the image formation. The radiation that falls on the neighboring pixels represents the energy lost to the image-forming pixel, while it is added as the optical noise to the neighboring pixels [11]. This formulation includes the consideration that the pixels have a rectangular shape.

Next, we consider the absorption of the energy by a signal-carrying pixel. We may write Φ_P in Equation (2) as a product of relative power φ_P and the *EOD*, introduced in Equation (1).

$$\Phi_P(d_x, d_y) = EOD(d_x, d_y)\ \varphi_P = \varphi_P \frac{\int_{-d_y}^{d_y}\int_{-d_x}^{d_x} psf(x,y)dx\,dy}{\int_{-\infty}^{\infty}\int_{-\infty}^{\infty} psf(x,y)dx\,dy\infty} \quad \left[\mathrm{W/cm^2}\right] \tag{3}$$

Using this expression, Equation (2) becomes

$$E_P(d_x, d_y) = \tau\varphi_P A_S\, EOD(d_x, d_y) = \tau\varphi_P A_S \frac{\int_{-d_y}^{d_y}\int_{-d_x}^{d_x} psf(x,y)dx\,dy}{\int_{-\infty}^{\infty}\int_{-\infty}^{\infty} psf(x,y)dx\,dy\infty}\ [\mathrm{W}] \tag{4}$$

This power, referred colloquially to as energy, has traditionally been referred to as "encircled", when the integration limit is the radial distance from the origin, or "enclosed" energy [12–21]. The maximum energy on a rectangular pixel, or the *EOD*, is obtained when the image centroid coincides with the center of the detector pixel. We denote this energy as the energy-on-detector pixel $E_d(d_x, d_y)$.

$$E_d(d_x, d_y) = max\left[\tau\varphi_P A_S\, EOD(d_x, d_y)\right] = max\left[\tau\varphi_P A_S \frac{\int_{-d_y}^{d_y}\int_{-d_x}^{d_x} psf(x,y)dx\,dy}{\int_{-\infty}^{\infty}\int_{-\infty}^{\infty} psf(x,y)dx\,dy}\right] \quad [\mathrm{W}] \tag{5}$$

An optical system afflicted with coma is an obvious case of an instrument in which the image centroid is displaced from its geometric image. In systems without coma, however, the image centroid and the center of the geometric image coincide.

The modern astronomical, space, or ground remote sensing instruments and telescopes, equipped with a large *FPA*, feature the state-of-the-art line-of-sight and pointing-control systems. Nevertheless, in a realistic environment with many sources of random motion, the image of a point source does not necessarily come to focus on the center of a desired detector pixel. This model was reasonable in the past, prior to pixelated detection schemes. The practitioners accept, to some degree, the unavoidable randomness in the pixel position [22–24].

The possibility of the displacement of the optical axis to an off axis point on the pixel may be described as a probability. We call this probability the optical centroiding efficiency, the *OCE*. We insert it as a correction factor into Equation (5) next to the *EOD* to obtain an accurate estimate of the radiative power transfer from a point source onto a detector pixel in a modern focal plane architecture.

$$E_d(d_x, d_y; \Delta d_x, \Delta d_y) = \tau\Phi_P A_S\ \left[EOD(d_x, d_y)\right]\left[OCE(\Delta d_x, \Delta d_y)\right] \quad [\mathrm{W}] \tag{6}$$

The need for such a correction factor has occasionally been accepted implicitly in earlier publications, employing a somewhat different terminology. We ascribe this unintentional historical oversight to the difficulty associated with its estimate for conditions of random environmental disturbances. The product *EOD* × *OCE* must be included in important radiometric concepts, such as the *NEP* and the figures of merit derived from it.

A realistic formula for the *NEP* of the point-source imaging sensor may then be reformulated, recalling its definition [25].

$$NEP = \frac{1}{\left[EOD(d_x, d_y)\right]\left[OCE(\Delta d_x, \Delta d_y)\right]}\frac{\sigma}{R} \quad [\mathrm{C/W}] \tag{7}$$

Here, σ is the detector noise in [counts] and R is the sensor responsivity in [counts/W] or [counts/(#photon s)]; C denotes counts.

3.2. Instrument Characteristic Function

The instrument characteristic function, or a point response function, *prf*, of a linear system is the convolution of the individual point response functions of the constituent subsystems [6]. The *prf(x,y)* of an optical instrument is the 2-D convolution of the *psf(x,y)*, the point spread function of the optical system (the imaging system of the camera or the telescope), with the detector response function *drf(x,y)*.

$$prf(x,y) = psf(x,y) * *drf(x,y) \tag{8}$$

The computer-assisted design, analysis, and optimization of optical systems has progressed to such an advanced degree that that *psf(x,y)* may be considered known. The magnitude of its Fourier transform, the modulation transfer function, is employed as one of the basic figures of merit for its performance evaluation. For the rectangular pixel, the detector (considered *spatial* here) response function, *drf(x,y)*, has the usual form,

$$drf(x,y) = \begin{cases} 1 & \begin{cases} -d_x < x < d_x \\ -d_y < y < d_y \end{cases} \\ 0 & elsewhere \end{cases} . \quad \left[\mathrm{cm}^2\right] \tag{9}$$

We may also consider *drf(x,y)* as the pixel aperture function. We use $2d_x$ by $2d_y$ as pixel dimensions. This function is sometimes denoted as *rect(x,y)*. It has the property that its one-dimensional Fourier transform is a *sinc(x)* = *sinx/x*. The pixel center is, in general, not aligned with the axis of the optical system; the pixel center is displaced randomly, in direction and magnitude, from the optical axis of the image-forming system. The image centroid falls on point (Δd_x, Δd_y). We consider that the coordinates of the detector-pixel center define the origin of the coordinate system, because the detector-pixel geometry introduces a fixed reference in the image detection and readout.

Using Equation (8) and displaced $psf(x{-}\Delta d_x, y{-}\Delta d_y)$, we obtain the instrument point response function, *prf*, or the instrument function,

$$prf(x,y;\Delta d_x,\Delta d_y) = psf(x-\Delta d_x, y-\Delta d_y) * *rect\left(\frac{x}{2d_x}, \frac{y}{2d_y}\right) \tag{10}$$

As a reference, the *psf(x,y)* of a perfect optical system with a circular aperture is a Bessel function, presented in Figure 2a. Using the definition of convolution, we find

$$prf(x,y;\Delta d_x,\Delta d_y) = \int_{-\infty}^{\infty}\int_{-\infty}^{\infty} psf(s-\Delta d_x, t-\Delta d_y)rect\left(\frac{x-s}{2d_x}, \frac{y-t}{2d_y}\right)dsdt. \tag{11}$$

We could limit the spatial extend of the *PSF* (the Fourier Transform of *psf*) to 10-pixel sizes, meaning that the integration limits are equal to $+/-10\mathrm{d_x}$, $+/-10\mathrm{d_y}$ (see also Figure 1).

$$prf(x,y;\Delta d_x,\Delta d_y) = \int_{-10d_y}^{10d_y}\int_{-10d_x}^{10d_x} psf(s-\Delta d_x, t-\Delta d_y)rect\left(\frac{x-s}{2d_x}, \frac{y-t}{2d_y}\right)dsdt \tag{12}$$

The rect-function of the pixel dimensions just defines the integration limits to smaller values.

$$prf(x,y;\Delta d_x,\Delta d_y) = \int_{-d_y}^{d_y}\int_{-d_x}^{d_x} psf(v-x-\Delta d_x, w-y-\Delta d_y)dvdw \tag{13}$$

The instrument function is, then, the *psf*, displaced by the optical axis displacement, integrated for each value of (*v,w*), over the area of overlap of the *psf* function and the pixel extent. Its maximum value, for a given *psf(x,y)* of the optical system and the detector sides, is found when their coordinate axes coincide, that is, when Δd_x, Δd_y are zero.

The peak value of the $prf(x,y)$, the instrument response function, is the EOD of the point source, found when Δd_x, Δd_y are zero. Thus, recalling Equation (1) for the energy on detector, the EOD, we modify Equation (8), resulting in the normalized $prf_n(x,y)$. We denote the pixel area, $4d_xd_y$, with A_d.

$$prf_n(x,y) = \frac{psf(x,y) * *drf(x,y)}{A_d \int_{-\infty}^{\infty} \int_{-\infty}^{\infty} psf(x,y)dx\,dy} \tag{14}$$

The $EOD_d(x,y)$ is the ensquared energy, that is, the energy enclosed within a pixel of dimensions $2d_x$ by $2d_y$ when the image centroid is not coincident with the pixel center. In this equation (Δd_x, Δd_y) are the coordinates of the image centroid. Inserting the decentering in Equation (1), we have

$$EOD_d(d_x, d_y; \Delta d_x, \Delta d_y) = \frac{\int_{-d_y}^{d_y} \int_{-d_x}^{d_x} psf(x-\Delta d_x, y-\Delta d_y)dx\,dy}{\int_{-\infty}^{\infty} \int_{-\infty}^{\infty} psf(x,y)dx\,dy} \quad [\text{unitless}] \tag{15}$$

This is the energy on the detector pixel that is misaligned by (Δd_x,Δd_y) with respect to the optical axis. The OCE is, then, a single number that is defined as the ratio of the energy-on-detector pixel that is misaligned by (Δd_x,Δd_y) with respect to the optical axis, averaged over all possible misalignments, to the ensquared energy on the aligned detector pixel.

We treat the numerator of Equation (15) statistically, because of the lack of knowledge. By averaging over all possible displacements of the optical axis from the pixel center, we find the most likely value for the $OCE(d_x,d_y)$.

$$OCE(d_x, d_y) = \frac{\left(\frac{1}{A_d}\right) \int_{-d_y}^{d_y} \int_{-d_x}^{d_x} EOD_d(d_x, d_y; \Delta d_x, \Delta d_y)d(\Delta d_x)d(\Delta d_y)}{EOD(d_x, d_y)} \quad [\text{unitless}] \tag{16}$$

The $EOD_d(d_x,d_y;\Delta d_x,\Delta d_y)$ is given in Equation (15) and the $EOD(d_x,d_y)$ in Equation (1). The $OCE(d_x,d_y)$ is a statistical variable that depends only on the relationship between the pixel dimensions and the point spread function through $EOD_d(d_x,d_y;\Delta d_x,\Delta d_y)$ and $EOD(d_x,d_y)$.

The $OCE(d_x,d_y)$ is, therefore, defined as the normalized $EOD(d_x,d_y)$ average over a detector area, given the equal probability that the image centroid falls anywhere on the detector pixel. It may be computed when we know the $psf(x,y)$ of the optical system and the detector pixel size, $2d_x$ and $2d_y$. We are interested in the product of the OCE and the EOD because it enters the figures of merit of radiometric systems, as in Equation (7).

4. Modeling and Methods

Both the EOD and OCE depend on the point spread function $psf(x,y)$ of the optical system. The $psf(x,y)$ may be interpreted as the irradiance distribution of the image of a distant point source, normalized to 1 at the origin. In an imaging optical system, it may be found according to the wave diffraction theory of aberrations [26].

$$psf(x,y) = \frac{1}{(\lambda R)^4} \left| \int_{-\infty}^{\infty} \int_{-\infty}^{\infty} P(\xi,\eta) e^{-\pi i (x\xi + y\eta) D/\lambda R} d\xi d\eta \right|^2 \tag{17}$$

Here, λ is the wavelength of the radiation emitted by the point source and i is $(-1)^{1/2}$. D denotes the aperture diameter. Both ξ and η coordinates are assumed to have non-zero values in the image plane within the range $[(-d_x,+d_x), (-d_y,+d_y)]$. According to the formal analysis, the limits of integration range from minus infinity to plus infinity while the image is formed in the far field. R is the image distance. Furthermore, by the definition of the f-number, $F/\#$, $(\lambda R/D) = \lambda F/\#$, the dimensions in the image plane may be normalized by the system ($\lambda F/\#$)-product.

The pupil function $P(\xi,\eta)$ describes the pupil transmission characteristics. It is zero outside of the projected aperture and at the obstruction. Inside, it may be expressed as a wavefront aberration function $W(x,\eta)$ [27,28].

$$P(\xi,\eta) = A(\xi,\eta)e^{2\pi i W(\xi,\eta)/\lambda} \tag{18}$$

Here, $A(\xi,\eta)$ and $W(\xi,\eta)$ are the amplitude and the phase of the wavefront aberration function, respectively. For a perfect optical system with circular aperture and no aberration, the diffraction generates an image as a Bessel function of the first order. Its square, or normalized irradiance, is also known as the Airy function (see Figure 2a).

We used a simple and straightforward approach to compute both the $EOD(d_x,d_y)$ and the $OCE(d_x,d_y)$, presented schematically in Figure 6. We set up a lens in *CodeV* and calculated $psf(x,y)$ directly. We then exported $psf(x,y)$ to *MatLab*, where the *EOD* was verified one more time to ensure the consistency of numerical and ray-trace results. The *OCE* was obtained with *MatLab*, employing the formulas here derived. We modeled the detector pixel as a square with a dimension of *2d*. For the sake of due diligence, part of our work was also numerically evaluated (using Equations in Figure 6) with *Mathcad*. Both approaches agreed.

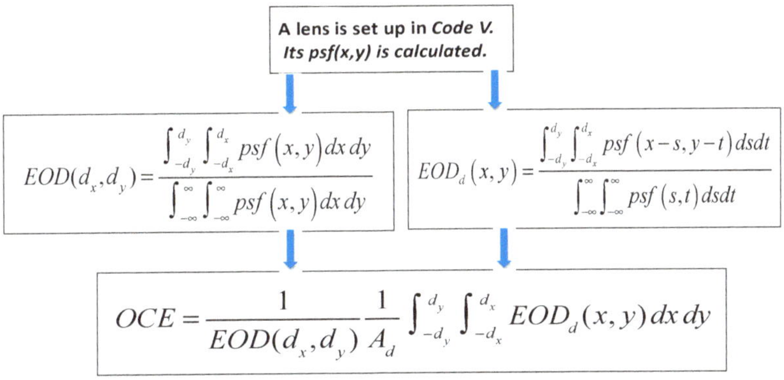

Figure 6. Block diagram to determine the *EOD* and *OCE*, using *CodeV* or some other lens design program to calculate the optical system $psf(x,y)$.

Next, we describe a study as an illustration of the process, related to the pixel dimensions. We discuss the determination of the OCE for an optical system with and without a generalized rectangular obscuration. The difficulties and the repetition inherent to a statistical analysis indicate why the concepts of energy on detector and optical centroiding efficiency have not been implemented previously. The implementation through the numerical evaluations are quite time consuming.

5. Effects of Pixel Size and Central Obscuration

We start this analysis with the detector size effects for an ideal optical system with and without a central obscuration. We examine the cases of a circular aperture with and without a rectangular central obscuration.

5.1. Aperture Configuration: No Central Obscuration

We analyzed two aperture types: an ideal circular aperture of diameter *D* and an ideal circular aperture with a rectangular central obscuration. The obscuration size was $0.27\ D \times 0.46\ D$, where *D* is the diameter of the entrance pupil (see Figure 5a).

We first analyzed the case of unobstructed aperture. We present its *OCE* as a function of detector size from $1.67\ \lambda F/\#$ to $15\ \lambda F/\#$ in Figure 7. We recall that the radius of the Airy disc was $1.22\ \lambda F/\#$. Thus, we started with the pixel size that did not quite enclose the central peak of the $psf(x,y)$ or the Airy disc, even when the image of the point source fell on the center of the pixel. For the larger detector sizes, the radius of the image of the central disc was increasingly smaller than the pixel dimensions. This figure shows that the *OCE* increased with the pixel size.

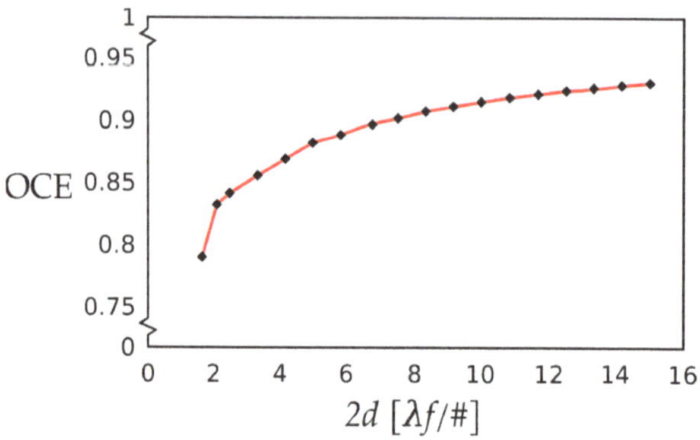

Figure 7. Optical centroid efficiency vs. detector pixel size in units of *[λF/#]* for a perfect optical system with a circular aperture, without a central obscuration.

From the block diagram presented on Figure 6, we noted that the construction of the *OCE* requires that the $EOD(d_x,d_y)$ and $EOD_d(d_x,d_y)$ be first determined. Therefore, we first calculated the *prf(x,y)*. Figure 8 presents the instrument point response function, *prf(x,y)*, vs. pixel number for two orthogonal directions and two different pixel sizes, *2d = 3.33 λF/#* in (a), and 2d = *13.33 λF/#* in (b).

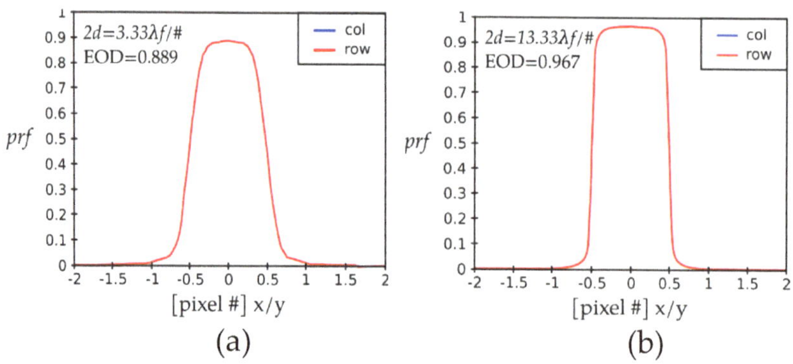

Figure 8. Two *prf*-s vs. pixel number shown for two orthogonal directions, x and y (in the inset denoted as red, rows, and blue, columns): (**a**) 2d = *3.33 λF/#* and (**b**) 2d = *13.33 λF/#*. Pixel coordinates along both dimensions are, in both cases, normalized to the pixel size. The general features of the curves for the small (**left**) and large (**right**) pixel sizes are quite similar, except that the peak for the smaller pixel (*3.33 λF/#*) is decreased by about 4% with respect to the peak of the large pixel. Additionally, the top of the profile for the smaller pixel size (**left**) is more rounded and narrower than that of the large pixel (**right**).

We next analyze a small detector pixel with a dimension of *3.33 λF/#* to present our methodology and to illustrate the salient steps.

5.1.1. Case 1: Small Pixel, No Central Obscuration

When the detector size was small, *3.33 λF/#*, the pixel comfortably enclosed the complete image of the point source. The diameter of the Airy disc (*2.44 λF/#*) was equal to about 2/3 of the pixel side. The convolution of the detector area with the Airy function started to produce ripples on both sides of the maximum of the *prf*-profile. More energy was distributed outside of the reference detector pixel, resulting in a reduced *OCE*. We present this case in Figure 8a, exhibiting the *prf(x,y)* cross-section as a function of pixel number for a *3.33* $\lambda F/\#$-pixel size. Rows and columns are superimposed because the optical system is rotationally symmetric.

We note that the peak portion looks like a flat top with rounded edges, with a peak value (that is, *EOD*) of *0.889*. The curve is broadened at the base. The *OCE* for this case was calculated by convolution and averaging to be *0.855*. The values (*3.33* $\lambda F/\#$, *0.855*) are entered as a single point in Figure 7. By neglecting the effects of the *OCE*, we overestimated the instrument performance. The product of *EOD* by *OCE* was *0.760*.

Next, we analyze the case of the detector pixel with dimension of *13.33* $\lambda F/\#$ to illustrate the effect of the pixel size.

5.1.2. Case 2: Large Pixel, No Central Obscuration

Figure 8b presents a *prf(x,y)* vs. pixel number for two orthogonal directions, x (red, row) and y (blue, column) for *13.33* $\lambda F/\#$ pixel size. The curves along the horizontal and vertical direction overlap because we analyzed a rotationally symmetric circular aperture and an ideal, diffraction-limited system. The detector physical extent ranges from -0.5 to $+0.5$ along the *x*- and *y*-axis. The pixel size of *13.33* $\lambda F/\#$ is approximately 5 times the diameter of the Airy disc; thus, the image of a point source is contained completely inside this pixel. The pixel-side dimension on the abscissa is normalized to 1.0 (in units of $\lambda F/\#$). The peak of the curve is centered at the origin because the square pixel intercepted the highest amount of radiation when the image was centered at the origin. Furthermore, because the pixel size is so much larger than the Airy disc diameter, the peak is nearly flat. This means that the image may be appreciably decentered over the pixel surface, nearly to the edge, before the *prf* changes.

The peak value of *prf* is the *EOD*. From Figure 8b, we read that $EOD(d_x,d_y)$ was *0.967* for the *13.33* $\lambda F/\#$-pixel size. The *psf* of the optical system was convolved with the detector pixel response function (*drf*) to find $EOD_d(d_x,d_y)$. Next, the *OCE* value was calculated by integrating the *prf(x,y)* over the detector area ($-0.5 < x < 0.5$, $-0.5 < y < 0.5$). We calculated a value of *0.926* for this *OCE*. The product of the *EOD by OCE* (0.967×0.926) was *0.895*.

The size of the Airy disc, or more precisely, the diameter of the first dark ring, was *2.44* $\lambda F/\#$. For a relatively large detector size of *13.33* $\lambda F/\#$, the resulting *prf(x,y)* presented a rather flat response similar to a hat, except for the slightly rounded corners, as seen in Figure 8b. These values (*13.33* $\lambda F/\#$, *0.926*) are entered as a single point in Figure 7. By neglecting the effects of the *OCE*, we overestimated the instrument performance.

In addition to the pixel size, the pupil geometry also plays an important role in the amount of radiation collected at the image pixel for a point object at infinity.

5.2. *Aperture Configuration: Circular, Incorporating Rectangular Central Obscuration*

The optical system features a rectangular central obscuration of Figure 5a. The obscuration size was $0.27\,D \times 0.46\,D$, where *D* is the diameter of the entrance pupil. The rectangular central obscuration breaks the circular symmetry, resulting in an increased relative intensity level in the first ring and decreased diameter of the Airy disc [27]. Figure 5b presents the point spread function, *psf(x,y)*, as a function of pixel number, in units of [$\lambda F/\#$], for two orthogonal directions, x (rows, red) and y (columns, blue).

Case 3: Small Pixel, Round Aperture with a Central Rectangular Obscuration

The pupil geometry-driven diffraction patterns incident on the pixel resulted in a rise in additional features in the shape of the *OCE* and the *EOD* curves. This may be observed in the OCE vs. detector size curves generated for the special case of an optical system with the central obscuration, exhibited in Figure 9. We modeled the case where the diameter of the first zero–zero ring was equal to 2 pixels (in units of $\lambda F/\#$). We observed a small negative spike in the small pixel-size regime, at about *3.4* $\lambda F/\#$. Afterwards, the curve resumed a trend, like that presented in Figure 7. With nearly 60 percent obstruction by area, this curve was expected to behave similarly to the lowest one in Figure 1. However, the rectangular pixel and rectangular obstruction were somewhat *in resonance* to produce a pronounced peak, followed by an unexpected dip in the small pixel-size regime.

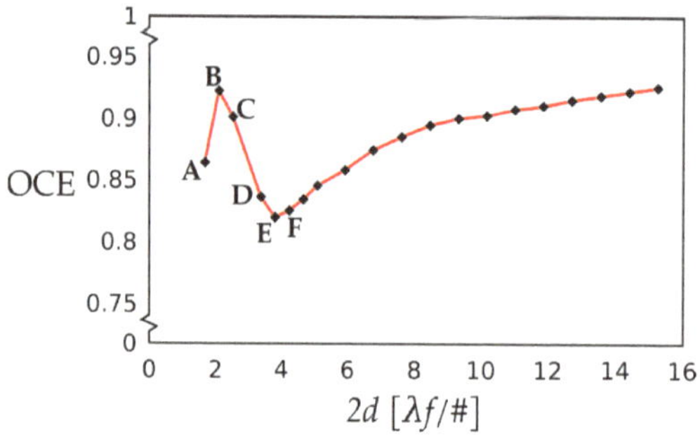

Figure 9. The *OCE* as a function of pixel size in [λF/#] for an optical system with a central rectangular obscuration (see Figure 5a). We observed a sharp increase in the *OCE* value toward the first peak to about *0.93* at about *2.05* pixels (Point B), followed by a sharp dip to about *0.82* (Point E) at about 4 pixels. Only from this, the lowest point on the curve, its shape starts to increase in a monotonical fashion.

We investigated further the first six data points in the unusual spike region where the detector size ranged from *1.67 λF/#* to *4.23 λF/#*. We followed the same analysis steps as we did in Cases 1 and 2. We plotted the corresponding *prf(x,y)* cross-section graphs for the six key points indicated in Figure 9 as parts (a) through (f) in Figure 10.

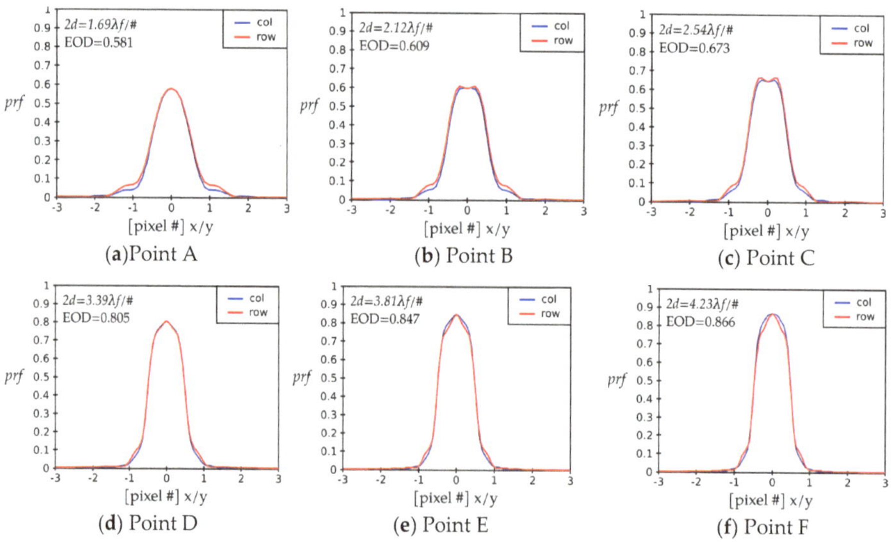

Figure 10. Cross-sections of the *prf* as a function of two orthogonal directions, x and y, or rows and columns, for the points labeled in Figure 9. We modeled the case where the diameter of the first zero in the ideal diffraction pattern was equal to 2 pixels (in λF/# units) with the rectangular obscuration of Figure 4a. We observed that the *EOD* value increased with increasing detector pixel size. *EOD* is the value of the *prf* at the pixel center. Concurrently, the shape of the *prf(x,y)* became modified with increasing detector size from that with explicit support at the base to that with a thin base and without features there.

The pixel size was equal to the diameter of the Airy disc just before Point C. When the pixel size was smaller than the disk diameter, the increase in the pixel size resulted in an increased energy on detector. Moving the beam center around the detector surface area did not affect the amount of collected energy, because the Airy function is relatively flat at the central part. Therefore, the overestimation in the collected energy for this region is quite significant.

Case 4, presenting a large pixel and circular aperture with central rectangular obscuration, does not warrant a separate analysis. The *OCE* on Figure 9 for large pixel sizes has a similar trend as that in Figure 7.

6. Discussion

Even when a sensor is equipped with a high-quality line-of-sight control, the centroid of the image of a point source falls somewhere between the center and the edge of a detector pixel for most realistic cases. This results in a decreased signal-carrying energy on detector.

6.1. Signal-Carrying Energy on Detector

We demonstrated that each point on the *prf* curve is the value of the energy on detector at that point when the image centroid of a point source comes into focus at that point. If we simply applied the calculated value of the *EOD* (of *0.967*) to characterize the performance of the FPA with a detector pitch of, for instance, *13.33* $\lambda F/\#$, this would imply that the centroid of each point source is always imaged at the pixel center. This assumption would generate an overly optimistic performance prediction for this FPA sensor because some energy is spilled on the neighboring pixels when the image of a point source is imaged at another position on the detector pixel. The more the image centroid is displaced from the pixel center, the more energy is spilled on the neighboring pixels and, correspondingly, less energy is contributed to the image formation.

Thus, a properly determined energy on detector used in the development of the detector and system figures of merit would have the peak *EOD* value modified by the factor of *OCE* which accounts for all possible misalignments between the optical system axis and the pixel axis. Including the calculated *OCE* of 0.926, the product of *EOD by OCE* is *0.895*. This is a more realistic figure of merit of the sensor performance. By neglecting the effects misalignment, and therefore those of the *OCE*, we overestimate the instrument performance by about 10% in the optical system with a circular aperture without a central obscuration.

In summary, comparing both sides of Figure 8, we may form the first significant conclusion for a diffraction-limited case with no central obscuration. As the detector pixel size increases relative to the image size, the *EOD* increases together with the associated *OCE*. By considering that the image of a point source falls on the pixel center, at least 10% energy is overestimated in performance measures.

6.2. Small, Medium, and Large Pixels to Collect Signal-Carrying Energy

We performed an illustrative study of the *OCE* vs. detector pixel size for a circular aperture with and without the rectangular central obscuration, displayed in Figure 11. We treated three regions of interest separately.

We first consider the case without the central obscuration. When the pixel size (*2d*) is smaller than the *psf* diameter, used in applications of shape recognition and object height contouring, the *EOD* increases together with the associated *OCE*. Additionally, the *EOD* increases according to the traditional concept of the enclosed energy. The *OCE*, however, increases at a much faster rate than the *EOD*, because *psf* is nearly constant over the pixel up to the pixel edge.

For the intermediate regions of pixels being equal to two to four central spots, the energy on detector still increases, but less rapidly than in the small pixel regime. The *OCT* increases in this region because the energy on detector is still relatively large at the pixel edge.

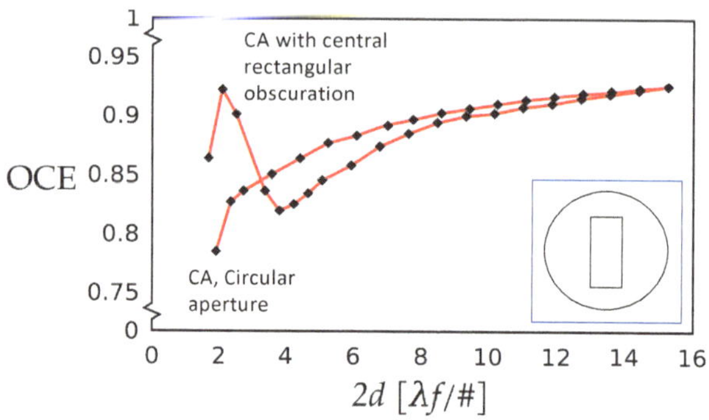

Figure 11. The *OCE* as a function of pixel size in [$\lambda F/\#$] for an optical system with and without a central rectangular obscuration. A sharp increase in the *OCE* value toward the first peak at about *2.05* pixels is followed by a sharp dip at about 4 pixels only for the case of the circular aperture with a central obscuration. The OCE curve for a circular aperture without a central obscuration increases in a monotonical fashion. CA denotes a circular aperture.

When the detector pixel size becomes larger than four central spots, the *EOD* increases start to moderate according to the traditional concept of the enclosed energy. The *OCE* increases slow down with the increase in the detector pixel size, because at large distances, there is increasingly small increment in detected energy. At the edge of a relatively large pixel, the *psf* is already small, resulting in a large fraction of the pixel area where the image may not be completely detected to produce a large *OCE*, approaching its limiting value of nearly 0.90.

The situation is appreciably different in the case of optical systems with a sizable central obscuration. They are characterized as having a smaller spot diameter, and more energy is spread into outer rings than those with a circular aperture without central obscuration. Thus, such a system could be required to have somewhat larger pixels to collect a slightly spread-out spot.

In the region where the pixel size is smaller than the central spot diameter, the *EOD* for a system with a large central obscuration increases with the pixel size more slowly than for a clear system. We already observed this phenomenon in the enclosed energy graphs in Figure 1. The *OCE*, however, increases very rapidly when the central spot is smaller than the pixel. The value of the sensor point response function *prf* at the pixel edge is still high just because the pixel is smaller than the central spot. Table 1 summarizes the point numbers, the pixel size, the *EOD* from Figure 9, the *OCE* values from Figure 10, and their product.

Table 1. The *EOD*, the *OCE*, and their product for small and medium pixel sizes and a circular aperture with a rectangular obscuration.

Point on Figure 9	*2d* [λF#]	*EOD* from Figure 10	*OCE* from Figure 9	*OCE* × *EOD*
A	1.69	0.581	0.864	0.502
B	2.12	0.609	0.922	0.561
C	2.54	0.673	0.901	0.606
D	3.39	0.805	0.836	0.673
E	3.81	0.847	0.820	0.694
F	4.23	0.866	0.825	0.714

When the pixel size is equal to about two to four spot diameters, the *psr* value at the pixel edge stars becoming increasingly smaller, close to zero. The *EOD* still increases with the detector pixel size because energy is spread out to higher-order rings nearly linearly (see Figure 1, bottom curve). The outer pixel regions, however, start to make a decreased contribution to the overall *OCE*, because the *prf* value there is low. In this region, we observe a pronounced decrease in the *OCE* as a function of the pixel size.

For pixel sizes equal to or larger than about four bright spots, the *OCE* starts to increase at a somewhat faster rate than in the case of the clear aperture. In this region, the *EOD* also increases at a rate that depends on the amount of obscuration. Yet, the side lobes have a lot of energy, so increasing the pixel dimension also results in a steady increase in the *OCE* values.

The energy on detector always increases with the detector size because a larger area of the detector always intercepts more energy.

Based on this analysis, we may conclude that the *OCE* is a meaningful quantity in the case when the detector pixel size is sufficiently large that it collects a significant fraction of energy in the image of a point source. In the case of a large obscuration with a transmission loss of about *0.6*, the smallest reasonable value for *EOD* is about *0.85*. This means that in diffraction-limited cases, the concept of the *OCE* is applicable and relevant for cases with relatively small central obscurations.

The *OCE* analysis does not apply to the cases where large amount of energy is spread out from the central spot, or where the detector is smaller than the central spot diameter.

The size of the pixel, which is smaller than the diameter of the Airy disc, corresponds to the case when imaging that requires high topographic details is performed, as in the Mars landing survey, and where the amount of incident energy is not an issue. Thus, one would one expect to analyze the displacement of the pixel and the optical system axes for such cases.

7. Summary and Conclusions

We introduced novel concepts and rationale for more appropriate terminology to describe the image detection process in modern digital instruments. The first one is the energy on detector (*EOD*), that replaces the old concept of energy enclosed inside a circle to assist with the performance measures of the pixelated sensors. We further argued that the image of a point source at infinity neither usually nor generally falls in the center of a pixel. Rather, in modern instruments, the axis of an optical system intersects a pixel at any point on the pixel surface. Considering that this point is unknown, we proposed as the second novel concept the optical centroiding efficiency (*OCE*). It is found by convolving the point spread function (*psf*) of the optical system with the detector responsivity function for all possible positions on the pixel. The product of these two quantities ($EOD \times OCE$) is proposed as a more realistic quantity to assess the sensor figures of merit.

Using the linear system theory in the instrument analysis, we developed a theory to evaluate the *EOD* and the *OCE*. We implemented it numerically on the *CodeV* and *MatLab*.

We performed two studies to illustrate the usefulness of these new concepts. (i) The *OCE* vs. detector pixel size for an ideal circular aperture. (ii) The *OCE* vs. detector pixel size for a circular aperture with a rectangular central obscuration. Based on this analysis, we may conclude that the *OCE* as a design parameter is a meaningful quantity in the case when the detector pixel size is sufficiently large that it collects a significant fraction of energy in the image of a point source. In the case of a large obscuration with a transmission loss of about *0.6*, the smallest reasonable value for the *EOD* is about *0.85*.

The *OCE* is proposed as the meaningful concept in the system performance assessment and in the development of figures of merit when the resolution of optical system is matched to the pixel size.

In the future, we want to extend this study to aberrations. We also plan to apply this theory to other telescopes and remote sensing instruments.

Author Contributions: All authors contributed equally to this M.S. Conceptualization, Y.W.; Investigation, R.M. and Y.W.; Methodology, R.M. and M.S.; Visualization, Y.W., M.S. and B.B.-M.; Writing—original draft, M.S. and Y.W.; Writing—review and editing, M.S. and B.B.-M. All authors have read and agreed to the published version of the manuscript.

Funding: This research received no external funding.

Institutional Review Board Statement: Not Applicable.

Informed Consent Statement: Not applicable.

Data Availability Statement: The data used in this study are included in this study.

Acknowledgments: This paper is based on a presentation prepared for the "1st International Conference–Advances in 3OM: Opto-Mechatronics, Opto-Mechanics, and Optical Metrology," 13–16 December 2021, Timisoara, Romania.

Conflicts of Interest: The authors declare no conflict of interest.

References

1. Rauscher, B.J.; Fox, O.; Ferruit, P.; Ferruit, R.J.; Waczynski, A.; Wen, Y.; Xia-Serafino, W.; Xia-Serafino, B.; Alexander, D.; Brambora, C.; et al. Detectors for the James Webb Space Telescope Near-Infrared Spectrograph I: Readout Mode, Noise Model, and Calibration Considerations. *arXiv* **2007**, arXiv:0706.2344. [CrossRef]
2. Sutherland, W.; Emerson, J.; Dalton, G.; Atad-Ettedgui, E.; Beard, S.; Bennett, B.; Bezawada, N.; Born, A.; Caldwell, M.; Clark, P.; et al. The Visible and Infrared Survey Telescope for Astronomy (VISTA): Design, technical overview, and performance. *Astron. Astrophys.* **2015**, *575*, A25. [CrossRef]
3. Scholl, M.S.; Wang, Y.; Randolph, J.E.; Ayon, J.A. Site certification imaging sensor for Mars exploration. *Opt. Eng.* **1991**, *30*, 590–597. [CrossRef]
4. Scholl, M.S.; Padila, G.; Wang, Y. Design of a high resolution telescope for an Imaging sensor to characterize a (Martian) landing-site. *Opt. Eng.* **1995**, *34*, 3222–3228. [CrossRef]
5. Scholl, M.S.; Padila, G. Push-broom reconnaissance camera with time expansion for a (Martian) landing-site certification. *Opt. Eng.* **1997**, *36*, 566–573. [CrossRef]
6. Goodman, J.W. *Introduction to Fourier Optics*; McGraw-Hill: San Francisco, CA, USA, 1968; p. 141.
7. Gaskill, J.D. *Linear Systems, Fourier Transforms, and Optics*; John Wiley & Sons: New York, NY, USA, 1978; p. 449.
8. Strojnik, M. Point spread function of (multiple) Bracewell interferometric configuration(s) and the nulling hypothesis in planet detection. *J. Appl. Remote Sens.* **2014**, *8*, 084981. [CrossRef]
9. Wolfe, W.L. Radiometric terms. In *The Infrared Handbook*, rev. ed.; IRIA, Ed.; ERIM for Office of Naval Research: Ann Arbor, MI, USA, 1993.
10. Wolfe, W.L. *Introduction to Infrared System Design*; SPIE: Bellingham, WA, USA, 1996; Volume TT24, p. 17.
11. Strojnik, M.; Scholl, M.K. Radiometry. In *Advanced Optical Instruments and Techniques*; Malacara, D., Thompso, B.N., Eds.; CRC Press: New York, NY, USA, 2018; pp. 459–717, ISBN 9781498720670.
12. Kingslake, R. *Lens Design Fundamentals*; Academic Press: New York, NY, USA, 1978; p. 153.
13. Fisher, R.; Tadic-Galeb, B. *Optical System Design*; McGraw-Hill: New York, NY, USA, 2000; p. 7.
14. Smith, W. *Modern Optical Engineering*, 4th ed.; McGraw Hill: New York, NY, USA, 2008; p. 191.
15. Hecht, E. *Optics*, 4th ed.; Addison Wesley: San Francisco, CA, USA, 2002; p. 470.
16. Mahajan, V. *Aberration Theory Made Simple*; SPIE: Bellingham, WA, USA, 1991; Volume TT6, p. 71.
17. Beyer, L.M.; Cobb, S.H.; Clune, L.C. Ensquared power for obscured circular pupils with off-center imaging. *Appl. Opt.* **1991**, *30*, 3569. [CrossRef] [PubMed]
18. Harvey, J.E.; Ftaclas, C. Diffraction effects of telescope secondary mirror spiders on varies image-quality criteria. *Appl. Opt.* **1995**, *34*, 6337. [CrossRef] [PubMed]
19. Barhydt, H. Effect of F/number and other parameters on performance of nearly BLIP search and surveillance systems. *Opt. Eng.* **1978**, *17*, SR-28.
20. Barhydt, H. *Figures of Merit for Infrared Sensors*; SPIE: Bellingham, WA, USA, 1979; Volume 197, p. 64.
21. Lloyd, J.M. Fundamentals of electro-optical imaging systems analysis. Electro-optical systems design, analysis, and testing. In *The Infrared & Electro-Optical Systems Handbook*; Accetta, J.S., Shumaker, D.L., Dudzik, M., Eds.; ERIM: Ann Arbor, MI, USA; SPIE: Bellingham, WA, USA, 1993; Volume 4, p. 18.
22. Holst, G. *Electro-Optical Imaging System Performance*, 3rd ed.; SPIE: Bellingham, WA, USA, 2002; p. 204.
23. Born, M.; Wolf, E. *Principles of Optics*, 7th ed.; Cambridge U. Press: Cambridge, UK, 1999; p. 443.
24. Hudson, R.D. *Infrared System Engineering*; Wiley & Son: New York, NY, USA, 1969; pp. 270–429.
25. Lloyd, J.M. *Thermal Imaging Systems*; Plenum Press: New York, NY, USA, 1975; pp. 11–166.
26. Stoltzmann, D.E. The perfect point spread function. In *Applied Optics and Optical Engineering*; Shannon, R.R., Wyant, J.C., Eds.; Academic Press: New York, NY, USA, 1983; Volume IX, pp. 111–148.

27. Wyant, J.C.; Creath, K. Basic wavefront aberration theory for optical metrology. In *Applied Optics and Optical Engineering*; Shannon, R.R., Wyant, J.C., Eds.; Academic Press: New York, NY, USA, 1992; Volume XI, pp. 1–53.
28. Welford, W.T. *Aberrations of the Symmetrical Optical System*; Academic Press: London, UK, 1974; p. 22.

photonics

MDPI

Article

Simulation-Based Considerations on the Rayleigh Criterion in Super-Resolution Techniques Based on Speckle Interferometry

Yasuhiko Arai * and Tong Chen

Department of Mechanical Engineering, Faculty of Engineering Science, Kansai University, 3-3-35, Yamate-cho, Suita 564-8680, Osaka, Japan
* Correspondence: arai@kansai-u.ac.jp

Abstract: This study sought to explain the physical phenomenon that eludes the constraints of the Rayleigh criterion in the microstructure observation method using speckle interferometry, for which super-resolution has been experimentally proven; the study was conducted using computer simulations. Separating the light from two light sources in close proximity, which exceeded the Rayleigh criterion under incoherent light, was believed to be impossible. The simulation results, however, showed that when coherent light is used, the separation of two close points is not necessarily impossible if the light phases between the two points are different. Furthermore, the resolution of microstructure observation techniques based on speckle interferometry was discussed. A new interpretation of the Rayleigh criterion in super-resolution techniques based on speckle interferometry was reported.

Keywords: Rayleigh criterion; super-resolution; coherent light source; speckle interferometry; computer simulation

1. Introduction

Super-resolution technology is an important observational technique that supports advances in biotechnology. When super-resolution technology is viewed from a broad viewpoint, it can be broadly divided into two categories: optical subjects that must be considered when collecting images of micro-objects that exceed the diffraction limit of the lens and subjects that must be considered when processing the collected images. The previous category is based on the Rayleigh criterion in observation optics, Abbe's theory of image formation, etc. [1,2]. The second category can be considered image processing techniques, such as the Lucy–Richardson method [3,4], which processes the collected images.

In these categories, this study discusses the problems of super-resolution, which has already been reported as a technique for observing microstructures beyond the diffraction limit, based on speckle interferometry.

Traditionally, microstructural observations have been performed using optical microscopy. However, it is widely acknowledged that optical microscopy cannot observe microstructures that exceed the diffraction limit of observation optics, as indicated by the Rayleigh criterion [1,2].

Several techniques have been proposed to avoid the Rayleigh criterion in order to achieve super-resolution. For example, in biotechnology, fluorescent proteins have recently been used to observe microstructures [5–15]. Specifically, new techniques, such as photoactivated localisation microscopy (PALM) and stimulated emission depletion (STED), have been developed to facilitate new biotechnological research. In addition, imaging of nanoscale objects has been achieved by bringing dielectric microspheres into contact with the subject [16,17]. Furthermore, new nanoscale observation techniques have been reported, such as superlens imaging [18] of objects several nanometres in size has been attempted.

Although image acquisition beyond the Rayleigh criterion is considered impossible in optical microscopy, if it were possible, it would be conceivable to capture moving images for extended periods as well as high-speed photographs of dynamically active living organisms in two dimensions. If such techniques can be developed, image-capturing techniques using optical microscopy will become an attractive technology to support developments in bio-research. Therefore, the development of such technology is eagerly awaited.

Recently, a technique for observing the shape of microstructures beyond the diffraction limit, which analyses the phase of light based on speckle interferometry [19–21], has been reported [22–26]. This new observation technique achieves super-resolution by detecting the phase distribution of light from the observed object, instead of processing only information from images captured as a light intensity distribution, as in conventional techniques. Super-resolution is obtained by analysing the information from the viewed object as a phase distribution of light using the speckle interferometry method.

In this method, based on Abbe's image theory [27], scattered light with many ray vectors is used as illumination light to increase the number of rays passing through the lens aperture. The phase change at the confocal on the imaging element on the surface under test is reconstructed in the computer as a two-dimensional phase distribution, and the surface shape of the object under test is observed as a three-dimensional shape distribution.

However, this technique based on speckle interferometry has been able to achieve, albeit experimentally, observations that exceed the Rayleigh criterion, which for many years was thought to be unreachable. However, there is no clear explanation as to what kind of physical phenomena enable this observation technique to exceed the Rayleigh criterion.

In this study, physical phenomena that elude the constraints of the Rayleigh criterion in microstructure observation methods using speckle interferometry were explained with computer simulations using COMSOL Multiphysics [25], which is capable of electromagnetic field simulation analysis.

This study clarifies that the Rayleigh criterion [2], which is based on the analysis of the intensity distribution of light assuming traditional incoherent light, must take into account the phase variation between nearby light sources when dealing with coherent light.

2. Materials and Methods

2.1. Techniques for Observing Microstructures beyond the Diffraction Limit Based on Speckle Interferometry

In the light of the microstructure observation technique used in this study, for example, it is assumed that the cross section of the measured object shown in Figure 1a can be defined as f(x).

Based on this assumption, when a lateral shift δx is given to the measurement object, as shown in Figure 1b, the shape displacement occurring at each measurement point can be defined as $f(x) - f(x + \delta x)$ from the speckle interferometric measurement method presented in a previous report [22]. The displacement of the shape is then accurately measured using speckle interferometry, and the pseudo-differential value $\{f(x) - f(x + \delta x)\}/\delta x$ in the shift direction with respect to the shape is obtained by dividing the detected displacement by the lateral shift value. Furthermore, the shape of the measurement object can be reconstructed by integrating pseudo-differential values.

In the calculation process, the phase distribution obtained by integration is aligned in two dimensions based on the relationship between the positions of each confocal point (P'c) at each measurement point (Pc), as shown in Figure 1c, resulting in the reconstruction of a three-dimensional shape f(x).

If the Rayleigh criterion is exceeded at nearby measurement points, the microstructure cannot be observed according to the traditional idea. However, super-resolution has been experimentally performed using this method based on speckle interferometry under beyond the diffraction limit. Simulations were performed in this study to answer this question.

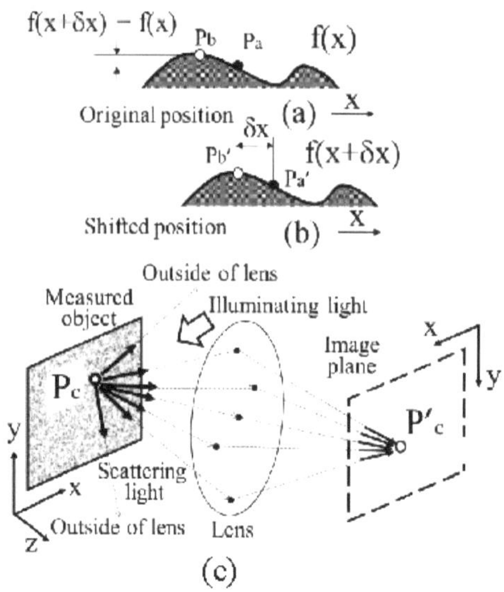

Figure 1. Principle of the measurement system. (**a**) Section of the measured object at the original position. (**b**) Section of the measured object at the shifted position. (**c**) Detection of two-dimensional phase distribution using the perfect optical system.

2.2. Simulation Model

As shown in a previous report [26], a speckle interferometer was constructed using a laser source with a wavelength of 532 nm, an objective lens (Mitutoyo M Plan Apo 200×) with a magnification of 200×, and an aperture (NA) of 0.62.

In speckle interferometry, only two speckle patterns are captured before and after the lateral shift of the measured object. Super-resolution images can be produced at a resolution of several tens of nanometers [23,24] using the speckle patterns.

However, it is extremely difficult to remove disturbances completely, such as the effects of stray light on the actual optical system, to confirm the principle of this method, as attempted in this study. In addition, it is difficult to discuss physical phenomena in detail owing to the limitations of measurement accuracy and the experimental environment.

This study investigates how electromagnetic simulation software (COMSOL Multiphysics) [25] can be used to observe microstructures beyond the diffraction limit.

The computer simulation model used in the study is shown in Figure 2a.

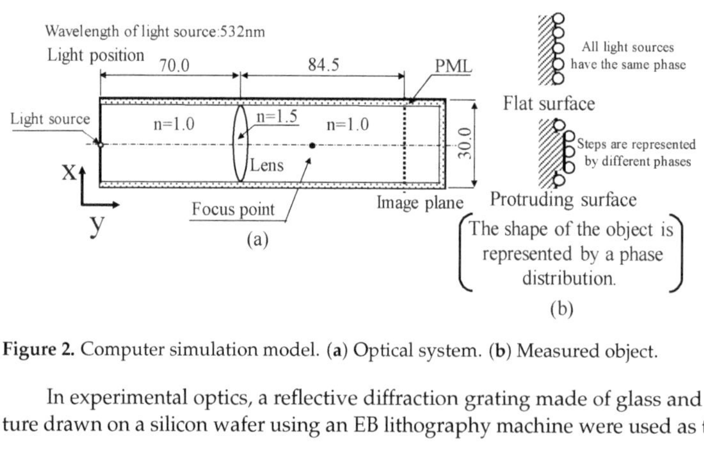

Figure 2. Computer simulation model. (**a**) Optical system. (**b**) Measured object.

In experimental optics, a reflective diffraction grating made of glass and a microstructure drawn on a silicon wafer using an EB lithography machine were used as the measured objects. The light reflected from the measured object [22–26] was analysed. The simulation described in a previous paper [25] set up a sinusoidal shape on the glass surface as the measurement object model. However, because this study is concerned with the Rayleigh criterion, the measurement objects were further simplified.

In other words, when a flat plane is the measurement object, as shown in Figure 2b, spherical wave light sources with the same phase are assumed to be lined up on the same surface.

When a surface with protrusions or steps is the object to be measured, the spherical wave light sources are arranged with the phase given according to the shape of the object by setting the wavelength of the light source to 2π rad so that it corresponds to the height of the protrusions.

In this way, the measurement object is set up, with the projection shape expressed as a phase distribution.

In the simulation model shown in Figure 2a, the mesh size was set to 1/12 of the wavelength, which was confirmed to have no effect on the calculation results even if the mesh size was not chopped any finer, in order to set conditions where the mesh size does not affect the calculation results as much as possible, while considering the load on the computer memory. As a result, it was confirmed that even when the mesh size was set to 1/12 of the wavelength, the difference in calculation results did not change more than 1% from the results with a mesh size of 1/20. In addition, the arrangement of light sources as measurement objects was also set with 0.4 nm as the minimum unit interval, while considering the load on the computer's memory. To effectively use the limited memory available, the parallel side walls and the right-side wall of the computational domain were defined as perfectly matched layers (PMLs). The simulation model was designed to minimise the load on the memory capacity by defining the computational domain with the minimum possible memory size (2 TB), as in a previous study [25].

The light sources used in the simulations were plane-wave and spherical-wave light sources defined by Equations (1) and (2), which are derived from Maxwell's equation [2]

$$E_P = Va \times \exp\{i(2py/\lambda + \phi)\} \quad (1)$$

$$Es = Va \times \frac{\exp\left\{i(2p/\lambda\sqrt{x^2+y^2}+\phi)\right\}}{\sqrt{x^2+y^2}} \quad (2)$$

In this study, the spherical wave source was used as the model for the scattered light used in speckle interferometry. In the light source model, Va is the electromagnetic field intensity, λ is the wavelength, and ϕ is the initial phase of light from the source. As described before, the phase distribution was used to set the shape of the measurement object.

The simulation model assumed the objective used is a thin biconvex lens; the refractive indices of air and the lens were defined as 1.0 and 1.5, respectively. Furthermore, as shown in Figure 3a, the focal length (f = 37.8 μm) was specified by determining the lens focal point as the point where the highest electromagnetic field intensity is focused by the lens when plane waves as collimated light are irradiated from the left wall surface to the lens. Furthermore, when the spherical wave source (Pd) is positioned on the optical axis of the left wall surface, as shown in Figure 3b, and the lens is positioned at a distance from the left wall surface by the focal length of the lens defined in Figure 3a, the electromagnetic field intensity after passing through the lens is confirmed to be collimated light.

From these results, the focal length f of this optical system was confirmed to be 37.8 μm. In general, the lens used here is designed with a glass with a refractive index of 1.5 by means of arcs with a radius of 40 μm. Since both convex surfaces of the lens are formed by arcs of radius 40 μm, the focal length can be obtained as 40 μm if the thickness of the lens is sufficiently thin [1,2]. However, since the thickness of the lens is not necessarily thin enough, 5.83 μm in relation to the lens diameter, the focal length in this study was determined using the procedure shown in Figure 3. As a result, the NA of the objective lens could be estimated as 0.37 [= 1 × sin (tan^{-1} (15/37.8))]. The diffraction limit as a Rayleigh criterion could then be obtained as 877 nm (=0.61 × λ/NA = 0.61 × 532/0.37).

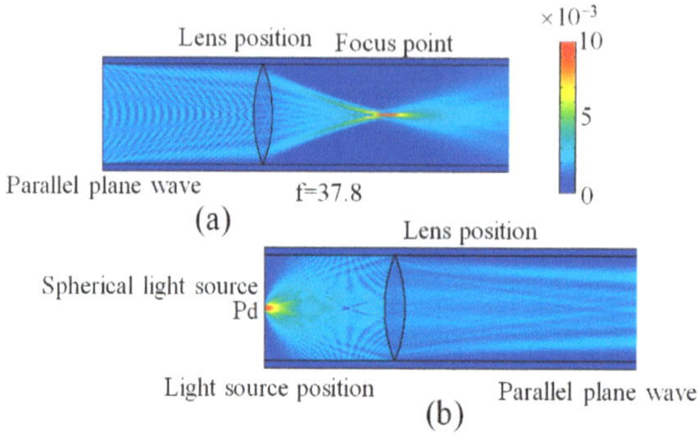

Figure 3. Focus length of the optical model. (**a**) Checking the focal point in the case of collimated light incidence. (**b**) Confirmation of collimated light at the focal point light source.

To further investigate the characteristics of the optical system, as shown in Figure 4, a spherical wave light source Pd (wavelength 532 nm) shown in Figure 3b was placed on the optical axis on the left wall of the optical system, and the focal length of the lens was set to 37.8 μm, as calculated in Figure 3.

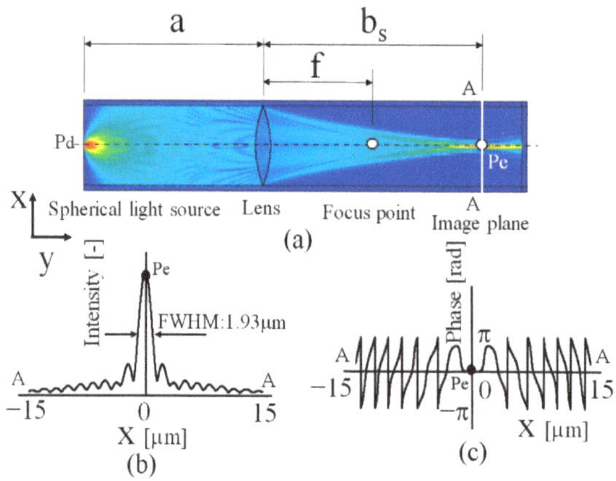

Figure 4. Optical system check when a light source is set on the light axis. (**a**) Electromagnetic field simulation result. (**b**) Intensity distribution in section A-A. (**c**) Phase distribution in section A-A.

The distance a between the lens and the light source and the distance b_s between the lens and the point Pe where the electromagnetic field intensity is the highest near the image formation position after passing through the lens were examined based on the Gaussian lens formula [1,2].

In this study, the distance b_s between the lens and the point Pe where the electromagnetic field intensity is the highest near the image formation position after passing through the lens was compared with the distance b_c obtained based on the lens formula. The results calculated using the values of the distance a to the light source and the focal length f of the lens yielded the values as b_c shown in Table 1.

Table 1. Factors of the optical system (μm).

a	f	b_c (Calculation)	b_s (Simulation)	b_c–b_s
60	37.8	102.2	105.8	−3.6
70	37.8	82.2	84.5	−2.3
80	37.8	71.7	74.7	−3

Comparing the value calculated based on the lens formula [1,2] with the distance between the point of highest intensity in the electromagnetic field intensity distribution obtained with computer simulation and the distance to the lens (b_s), it was confirmed that the values of b_c and b_s are approximately similar when a is set to 60, 70, or 80 μm, considering the thickness of the lens. Based on these results, the computer simulation model set up in this study is considered to reproduce the actual optical system. Based on the results of the study, a focal length of 37.8 μm was used in the following simulation model. Furthermore, by setting the distance between the measurement object and the lens to 70 μm, the simulation was conducted using the b_s (84.5 μm) values in Table 1 for the image formation position of the illumination light source. In this case, the lens magnification was 1.2×. Since this is a computer simulation, the phase distribution can be easily calculated using not only the intensity distribution of the light but also the real and imaginary parts of the intensity distribution.

In this case, the intensity distribution on the image plane (A-A in Figure 4a) is shown in Figure 4b and the phase distribution in Figure 4c.

On the imaging plane, an intensity distribution symmetrical in the x direction (Figure 4b) with a peak (Pe) on the optical axis can be confirmed. However, the phase distribution is obtained in the range of $-\pi$ to π rad, since the calculation result as a simulation result is not phase-unwrapped as an inverse tangent function of the ratio of the real and imaginary parts of the intensity distribution. It can be confirmed that the phase is 0 rad at point Pe on the imaging plane, which is considered confocal for a spherical light wave source with an initial phase of 0 rad.

Using this computer simulation model that identifies the fundamental properties of the optical system, this study examines the physical effects of the Rayleigh criterion on the measurement results in a super-resolution technique based on speckle interferometry.

3. Results and Discussion

3.1. Consideration of Rayleigh Criterion in Super-Resolution Technology Based on Speckle Interferometry by Simulation

3.1.1. Consideration of the Case Where the Two Light Sources do Not Exceed the Rayleigh Criterion

The spherical wave light source located on the optical axis on the left wall of the model shown in Figure 4a was newly replaced on the left wall as a spherical wave light source symmetrical to the optical axis separated by 2 μm across the optical axis with the same phase. The electromagnetic field intensity distribution when light is emitted from the two light sources is shown in Figure 5a. The intensity and phase distributions on the imaging plane in this case are shown in Figure 5b,c.

The distance between the light sources was 2 μm, and the diffraction limit of the optics was 877 nm, which means that the two light sources are set at positions that do not exceed the Rayleigh criterion. As a result, spherical wave beams from two points 2 μm apart interfere, and Young's fringes [1,2] are formed in the intensity distribution, as is generally well known. In the B-B section of the imaging plane, as shown in Figure 5b, although the intensity distributions are not completely separated, the two peaks can be observed because they do not exceed the Rayleigh criterion. It can also be clearly observed that the zeroth-order and ±first-order light of the Young's fringes pass through the lens aperture and are focused at the image formation plane. The phase distribution on the imaging plane in this case is shown in Figure 5c.

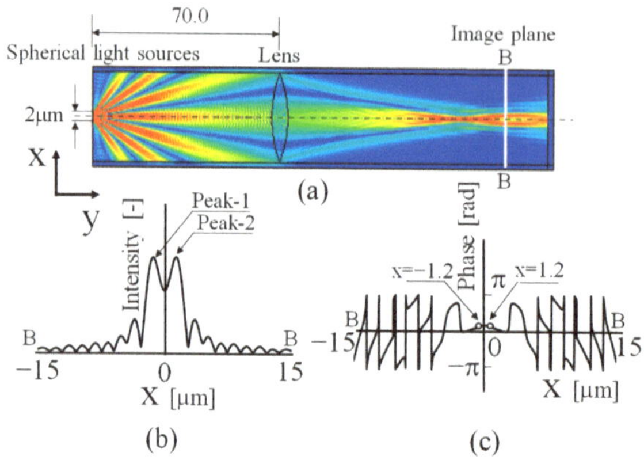

Figure 5. Simulated results when two light sources are set further than the Rayleigh criterion. (**a**) Electromagnetic field simulation result. (**b**) Intensity distribution in section B-B. (**c**) Phase distribution in section B-B.

When looking at the nearby optical axis in Figure 5c, the phase difference between the two light sources at x = −1.2 and x = 1.2 as the confocal where the light sources are set up was 0 rad, because the initial phases of the two spherical wave light sources set up in Figure 5a were both 0 rad.

That is, light from two spherical wave light sources with the same phase set at a distance that does not exceed the Rayleigh criterion can be considered not only as two light sources in terms of the intensity distribution on the image plane but also as the same phase in terms of phase distribution. This means that when considering the Rayleigh criterion in super-resolution technology based on speckle interferometry using coherent light as a light source, it is necessary not only to discuss the intensity distribution but also to investigate the phase distribution in detail, which has not been sufficiently investigated in the past.

3.1.2. Consideration of the Case where Light Sources Are Located at a Proximity Distance Exceeding the Rayleigh Criterion

Next, the case when the Rayleigh criterion in the earlier section is not exceeded occurred, and the distance between two spherical wave sources decreased from 2 μm specified in Figure 5 to 0.5 μm, as illustrated in Figure 6a.

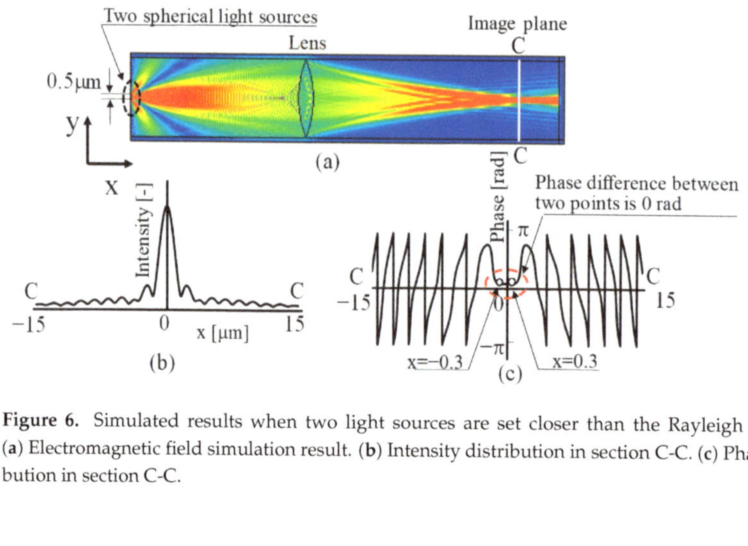

Figure 6. Simulated results when two light sources are set closer than the Rayleigh criterion. (**a**) Electromagnetic field simulation result. (**b**) Intensity distribution in section C-C. (**c**) Phase distribution in section C-C.

Figure 6a shows that as the distance between the two light sources becomes closer, the ±first-order light of the Young's fringes spreads out greatly to the left and right, clearly not passing through the lens aperture, and only the zeroth-order light passes through the lens and is focused at the image formation position. As a result, the intensity distribution on the imaging plane when emitting light from two points simultaneously, as shown in Figure 6b, cannot be observed as two separate light sources, even though the light was emitted separately from each of the two points. Instead, it is observed as a single peak with a maximum value near the centre between the peaks of the separately emitted lights. That is, the well-known phenomenon based on the Rayleigh criterion can be observed [1,2].

In this case, since the initial phase of the light source placed as a spherical wave source was set at the same value as 0 rad for both light sources, the phase difference of the observed light was 0 rad at the two points $x = -0.3$ and $x = 0.3$ close to the nearby axis surrounded by the red dashed line, as shown in Figure 6c.

It is considered that the phase difference corresponding to the initial phase of the two light sources is detected between the confocal points of the two light sources installed as the measurement object, as shown in Figure 5, on the image formation plane of the two light sources placed at a distance closer than the Rayleigh criterion. This means that when the diffraction limit based on the Rayleigh criterion is exceeded, the phase of the two light sources can be detected as the phase difference between the two points, although the intensity distribution can only be confirmed as a single point due to diffraction phenomena.

Thus, it can be understood that according to the traditional Rayleigh-criterion-based approach, imaging the shape of a measured object structure is a process based on intensity distribution, and therefore, due to phenomena caused by diffraction, it is not possible to observe microstructures beyond the diffraction limit using imaging techniques.

However, by treating the phase distribution and especially the phase difference between two light sources, it is possible to analyse the phase of light from each point, even if they are two points in close proximity, and there is a possibility that the shape of the measurement object can be reconstructed.

3.1.3. Consideration of Different Initial Phases of Light Sources Located at Close Proximity Distances Exceeding the Rayleigh Criterion

The difference between the experimental conditions based on super-resolution technology based on speckle interferometry and the simulation conditions when dealing with light from light sources of the same phase, as described in the previous section, is discussed next.

In the super-resolution technique based on speckle interferometry, when observing a microstructure, reflected light with a different phase is reflected from each point on the surface of the measured object, depending on the shape of the microstructure, and this reflected light is analysed.

However, the simulation in the previous section differs in that the light sources in close proximity have the same phase.

In this study, it was considered that the reflected light with different phases plays an important role in realising high resolution beyond the Rayleigh criterion in super-resolution technology based on the speckle interferometry technique.

Therefore, different from Figure 6, the initial phases of the two light sources were set as 0 rad and π rad, and the phase on the image formation plane was examined next when the distance between two points was set at 0.5 μm, as in Figure 6. The results are shown in Figure 7.

Comparing the electromagnetic field intensity distribution in Figure 7a with the result in Figure 6a, it can be seen that the phase of the intensity distribution reversed and the intensity of light near the optical axis weakened.

It can also be seen that the intensity of light in the diagonal directions, where the existence of intensity could not be observed in Figure 6a, became stronger. Since the initial phase differs by π rad between the two light sources, it is a natural result that the zeroth-order and ±first-order phases of Young's fringes in Figure 6a change by π rad.

As a result, light with strong intensity in the space between the zeroth- and ±first-order light in Figure 6a is considered to be generated, as shown in Figure 7a. In short, it can be understood that the separation of the two image points is due to destructive interference between the two images, as already suggested by microsphere-assisted microcopy [16,17].

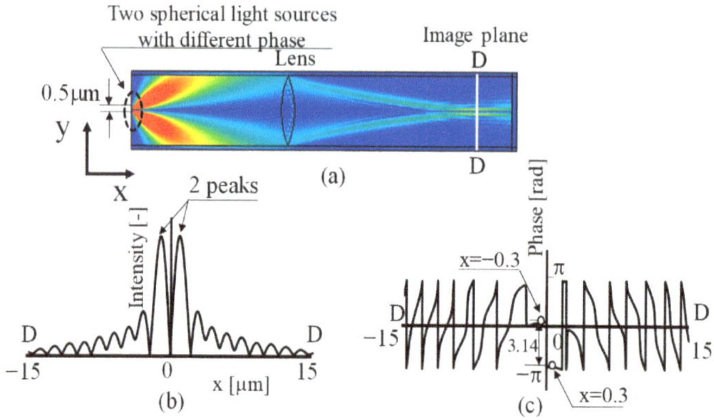

Figure 7. Simulated results when two light sources whose initial phases are different as π rad are set closer (0.5 µm) than the Rayleigh criterion. (**a**) Electromagnetic field simulation result. (**b**) Intensity distribution in section D-D. (**c**) Phase distribution in section D-D.

With this change in the direction of light, there is light that passes through the lens aperture, which did not exist in the in-phase case. This is thought to be the phenomenon of two bright spots on the image formation plane. In this case, when the light in this new direction is observed as an intensity distribution on the imaging plane, two intensity peaks can be observed near the optical axis, as shown in Figure 7b. Detailed observation of this phenomenon in Figure 7a shows that light emitted from the two points forms interference fringes known as Young's fringes and that the phase of the fringes is inverted and divided into two directions (upper and lower). Next, a part of the light from each of the two directions passes through the lens and reaches the image plane, forming two peaks as the intensity distribution. That is, Figure 7a,b confirms that there are two light points at the observation point on the image formation plane.

Furthermore, when observing the nearby optical axis of the phase distribution in Figure 7c in detail, the phase difference at the position of each white circle at x = −0.3 and x = 0.3, the confocal point of the two spherical wave light sources changed by π rad.

When the phase between the two light sources on the left wall differs by π rad, it can be confirmed that even if the two light sources are located beyond the Rayleigh criterion, the phase difference at the confocal point corresponding to the position of each light source set as the measurement object in the observed phase distribution differs by π rad. This means that even if the distance between two light sources exceeds the Rayleigh criterion, the phase difference between the light sources set up as light sources is preserved at the observation point.

This phenomenon suggests that the shape of an object can be measured beyond the diffraction limit as a phase distribution by detecting the phase at each point of the object with high resolution in super-resolution technology based on speckle interferometry technology. It can then be understood that for a phase distribution to exist, the existence of a geometrical unevenness distribution on the measured surface is required.

3.2. Experimental Verification of a Phenomenon Obtained in Simulation Results That Occurs Based on a Phase Change between Two Light Sources Located beyond the Diffraction Limit

In the simulation, it was shown that when the phases of two light sources in close proximity beyond the diffraction limit are different, the existence of the two light sources can be confirmed by detecting the phase difference between the two light sources if the light is coherent. In other words, if two light sources are based on the diffraction limit, which was thought to be impossible to confirm the existence of two points that exist beyond the Rayleigh criterion, based on the simulation results, the separation of the existence of two light sources beyond the diffraction limit is considered possible by detecting the phase difference between the two points with high resolution. Therefore, it was experimentally verified whether the phenomena based on the simulation results could occur in reality using a real optical model that was simplified as much as possible.

In the optical system used in the experiment, the diffracted image shown in Figure 8a with a circular aperture, formed by a laser light source with a source wavelength of 532 nm, was used as the diffracted image model [2] when the light source was observed using a circular lens. Two diffraction image models were prepared with light emitted from the same laser and with no phase difference between the two diffraction image models, and the two models were superimposed so that they overlapped from the left and right.

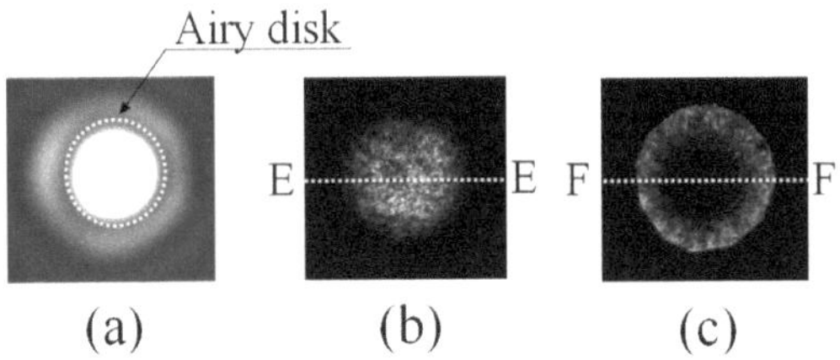

Figure 8. Overlap of diffracted lights with phase information. (**a**) Diffracted light source model. (**b**) Overlap of two light sources with phase difference 0 rad. (**c**) Overlap of two light sources with phase difference π rad.

In this case, if the overlap exceeds the Airy disk, the light source is considered only one light point. The general situation regarding diffraction limits based on the well-known Rayleigh criterion [2] arises. Furthermore, when the two lights are superimposed so that they gradually coincide spatially, the interior of the Airy disk is observed as a single bright light source because the phases of the two lights are originally equal, as shown in Figure 8b. This is also a well-known phenomenon that generally occurs when dealing with the Rayleigh criterion [2].

In this case, the E-E section of the intensity distribution in Figure 8b is shown in Figure 9a. It can be observed that the entire inner surface of the Airy disk is brightened.

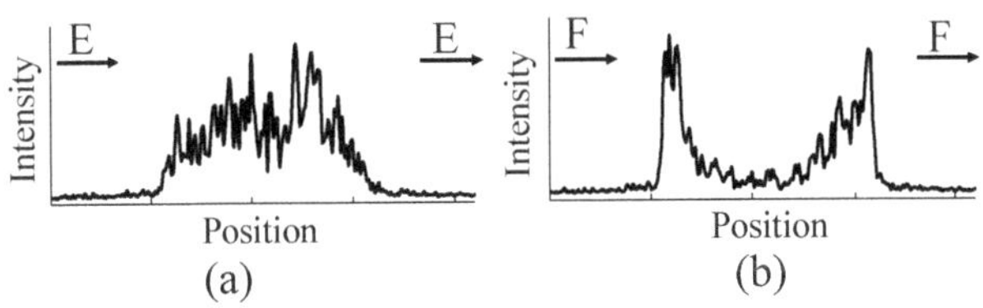

Figure 9. Section of intensity. (**a**) Two light sources with phase difference 0 rad. (**b**) Two light sources with phase difference π rad.

Next, the phase difference between the two overlapping lights was changed to π rad by changing the optical path length of one of the lights. It could be confirmed that the bright state inside the Airy disk, as shown in Figure 8b, changes to a dark state at the centre of the light, as shown in Figure 8c. In this case, the intensity distribution in the F-F cross section of Figure 8c is shown in Figure 9b. Clearly, the central area becomes darker and the peripheral area becomes brighter, just as in Figure 7b, which was observed in the simulation.

Thus, if the two light sources in close proximity beyond the Rayleigh criterion are coherent and the phases of the two light sources are different, experimental results confirm that the two light sources can be separated by detecting the phase difference between the two points on the image formation plane, as in the simulation results.

3.3. Influence on the Detection Phase of two Coherent Light Sources with Different Phases as the Distance between them Changes

As shown in this study, it was found that even two light sources exceeding the Rayleigh criterion can be observed as two points using phase analysis of coherent light.

In Figure 7, two light sources 0.5 μm apart were observed. Therefore, the next case in which the two light sources are even closer to each other was discussed.

The results for the case where the distance between the two light sources is 0.25 μm are shown in Figure 10. Figure 10a–c shows the results when the two light sources have the same phase (0 rad). Similar to the results shown in Figure 6, the two light sources placed beyond the diffraction limit cannot be separated. In the intensity distribution in Figure 10b, two light sources cannot be considered as two light sources. However, in the phase distribution in Figure 10c, it can be clearly confirmed that the phase difference between the two light sources is 0, as shown by the red dashed line.

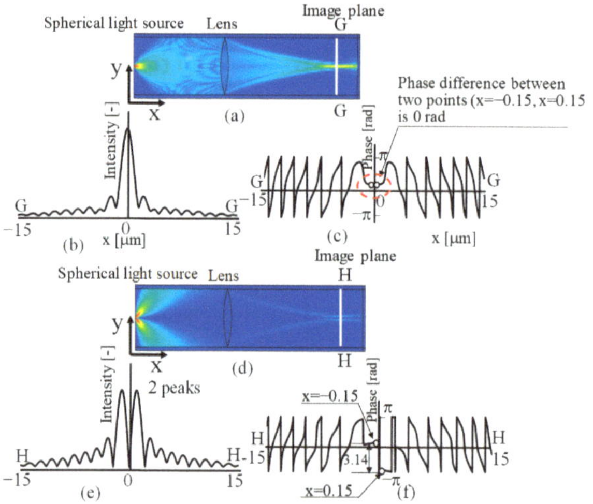

Figure 10. Simulated results with two light sources in close proximity to 0.25 μm, exceeding the Rayleigh criterion when the initial phases of the two light sources are in the same phase and differ by π rad. (**a**) Electromagnetic simulation results for the same phase. (**b**) Intensity distribution for the same phase in section G-G. (**c**) Phase distribution for the same phase in section G-G. (**d**) Electromagnetic field simulation results differ by π rad. (**e**) Intensity distributions differ by π rad in section H-H. (**f**) Phase distributions differ by π rad in section H-H.

In contrast, in Figure 10d–f, where the phase difference between the two light sources changes to π rad, two peaks can be observed in the intensity distribution in Figure 10e.

Furthermore, in the phase distribution in Figure 10f, it can be confirmed that the phase difference between the two points changes by π rad at the confocal point where the light source is located.

These results show that in an optical system with a diffraction limit of 877 nm, if two light sources 250 nm apart are coherent light sources and their phases are detected, it is possible to observe them as two light sources beyond the diffraction limit.

Furthermore, how an observation becomes possible when two light sources are in close proximity was investigated using simulation.

First, the light source wavelength was 532 nm, and the phase difference between the two light sources was set as π rad when the distance between the two light sources varied from 0.01 to 0.5 times the light source wavelength (λ).

In Figure 11, the horizontal axis is the distance between the two light sources (w) and is given as a multiple of the wavelength λ. The vertical axis is the detected phase difference between the two light sources.

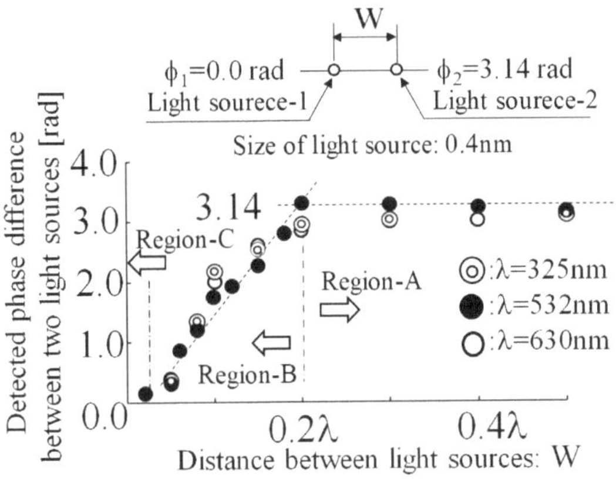

Figure 11. Relationship between the distance of light sources and the detected phase.

In the result for the wavelength of 532 nm indicated by the black circle (•), it can be observed that the phase of two light sources in close proximity can be detected as π rad, which is set as the phase difference, up to about 0.2 λ (actual length: about 100 nm). Next, as the distance becomes closer than 100 nm (0.2 λ), the phase difference gradually becomes increasingly smaller, and even if the phase difference is set as π rad, it can no longer be detected as π rad.

However, even if it is no longer possible to accurately detect dimensions related to the shape of the object, it is still possible to observe the approximate shape of the measured object. For example, in the experimental results of a previous report [26], although the groove depth of a 100-nm-wide groove could be detected almost accurately when observing a 100-nm-wide groove, the groove depth of a 60-nm-wide groove could not be accurately detected as the actual groove depth, although it could be captured as a groove.

Thus, in the observation of microstructures using speckle interferometry, there are several measurement limits in the experimental measurement process, such as the range where dimensions can be accurately measured (e.g., Region-A in Figure 11), the range where dimensions cannot be accurately measured but shapes can be captured (e.g., Region-B in Figure 11), and the range where the state of the measurement is not yet clearly understood (e.g., Region-C in Figure 11).

It is thought that there are several levels of measurement limits.

Therefore, it is necessary to investigate in detail the regions below 0.2 λ in Figure 11 (Region-B and Region-C in Figure 11) in the future. Furthermore, based on the results of this study, it is also necessary to discuss the causes of why such regions occur. Based on the results of these further investigations, the measurement limits of this method should be considered in more detail.

In this study, simulations were performed on the basis of the experimental results already reported. As a result, the wavelength was considered 532 nm. However, as a general concept in optical measurement, it is important to know how a change in wavelength affects the measurement results. Therefore, next, a simulation was performed to see how a light source with a different wavelength, as well as 532 nm, affected the measurement results.

For the light source wavelength, it was decided to consider visible light lasers, which are commonly used for measurements. The case of a long wavelength of 630 nm, modelled after a He-Ne laser (wavelength: 632.8 nm), is indicated by a white circle (○) in Figure 11.

The results are also shown in Figure 11 as double white circles (◎) for the case of the 325 nm wavelength modelled after the shorter wavelengths of He-Cd lasers (wavelengths: 325 nm and 442 nm).

For different light source wavelengths in Figure 11, it can be found that the phase difference between the two light sources can be accurately detected up to approximately one-fifth of the wavelength (0.2 λ), regardless of the wavelength. This indicates that even when the diffraction limit is exceeded, observation of finer structures becomes possible as the light source wavelength becomes shorter.

These results show that in the observation technique for structures with geometries beyond the diffraction limit using speckle interferometry, the phase difference between two nearby light sources is preserved during detection, even between two points beyond the diffraction limit, when coherent light is used.

By using this phenomenon, it is thought that super-resolution beyond the Rayleigh limit, which has been thought to be undetectable using the conventional Rayleigh criterion based on incoherent light, is realised by detecting the phase difference at each position in the microstructure observation technique based on speckle interferometry technology.

The discussion in this study also focused on super-resolution technology during image sampling, particularly for optical observation of microstructures. However, the sensing technology obtained in this study, which reveals the possibility of realising super-resolution based on phase manipulation of light waves with coherent properties, could also be applied to other sensing fields using electromagnetic waves, such as radar sensor technology [28].

In the future, the results of this research may lead to the use of phase manipulation technology in sensing related to super-resolution using electromagnetic waves with coherent properties, not only in the field of optics, but also in a wide range of other fields.

4. Conclusions

In this study, a physical explanation for the super-resolution phenomenon in a new microstructure observation technique using speckle interferometry [22], in which the realisation of super-resolution has been experimentally confirmed, was discussed. In this explanation, a computer simulation was used to investigate why the observation of microstructures exceeding the Rayleigh criterion, which had long been considered unexceedable, could be realised.

The simulation results show that when coherent light is used as a light source and the phase difference between two light sources is different, the phase difference is preserved at the image formation position at the time of detection, even if the two light sources are close to each other beyond the Rayleigh criterion.

By using this physical phenomenon, it was shown that light from two points in proximity exceeding the Rayleigh criterion can be detected as light from two points in proximity exceeding the Rayleigh criterion by capturing the phase distribution, although it was previously thought that light from two points in proximity exceeding the Rayleigh criterion cannot be separated on the basis of incoherent light.

Furthermore, regarding the measurement limit of super-resolution technology based on the speckle interferometry technique, the simulation model used in this study clarified that the phase difference between two light sources can be accurately detected up to a distance as close as about 20% of the light source wavelength. In the discussion that led to this conclusion, it became clear that this super-resolution technology has three types of measurement limits: (1) the range where dimensions can be accurately measured, (2) the range where dimensions cannot be accurately measured but shapes can be captured, and (3) the range where it is difficult to accurately measure differences in the object's steps, etc. In addition, it was confirmed that the discussion of these measurement limits is consistent with the results of the previous experiments.

Author Contributions: Conceptualization, Y.A.; methodology, Y.A.; software, Y.A. and T.C.; validation, Y.A.; formal analysis, Y.A and T.C.; investigation, Y.A. and T.C.; resources, Y.A.; data curation, Y.A. and T.C.; writing—original draft preparation, Y.A.; writing—review and editing, Y.A.; supervision, Y.A.; project administration, Y.A.; funding acquisition, Y.A. All authors have read and agreed to the published version of the manuscript.

Funding: This research was funded by JSPS KAKENHI (grant number 20H02165).

Institutional Review Board Statement: Not applicable.

Informed Consent Statement: Not applicable.

Data Availability Statement: Not applicable.

Acknowledgments: We would like to thank Dahai Mi of Keisoku Engineering System CO., LTD, for his kind guidance and support in the use of COMSOL Multiphysics.

Conflicts of Interest: The authors declare no conflict of interest.

References

1. Born, M.; Wolf, E. *Principles of Optics*, 7th ed.; Cambridge University Press: Cambridge, UK, 2019; pp. 167–177, 290–292, 370–374.
2. Hecht, E. *Optics*, 4th ed.; Addison-Wesley Publishing Co.: San Francisco, CA, USA, 2002; pp. 24–31, 149–165, 393–400, 467–474.
3. Lucy, L.B. An Iterative Technique for the Rectification of Observed Distributions. *Astron. J.* **1974**, *79*, 745–754. [CrossRef]
4. Richardson, W.H. Bayesian-based Iterative Method of Image Restoration. *J. Opt. Soc. Am.* **1972**, *62*, 55–59. [CrossRef]
5. Garini, Y.; Vermolen, B.J.; Young, I.T. From micro to nano: Recent advances in high-resolution microscopy. *Curr. Opin. Biotechnol.* **2005**, *16*, 3–12. [CrossRef]
6. Neice, A. Chapter 3—Methods and Limitations of Subwavelength Imaging. In *Advances in Imaging and Electron Physics*; Elsevier: Amsterdam, The Netherlands, 2010; Volume 163, pp. 117–140. [CrossRef]
7. Haeberle, O.; Belkebir, K.; Giovaninni, H.; Sentenac, A. Tomographic Diffractive Micros-copy: Basics, Techniques and Perspectives. *J. Mod. Opt.* **2010**, *57*, 686–699. [CrossRef]
8. Hell, S.W.; Wichmann, J. Breaking the diffraction resolution limit by stimulated emission: Stimulated-emission-depletion fluorescence microscopy. *Opt. Lett.* **1994**, *19*, 780–782. [CrossRef]
9. Klar, T.A.; Jakobs, S.; Dyba, M.; Egner, A.; Hell, S.W. Fluorescence microscopy with diffraction resolution barrier broken by stimulated emission. *Proc. Natl. Acad. Sci. USA* **2000**, *97*, 8206–8210. [CrossRef]
10. Heilemann, M.; van de Linde, S.; Schüttpelz, M.; Kasper, R.; Seefeldt, B.; Mukherjee, A.; Tinnefeld, P.; Sauer, M. Subdiffraction-resolution fluorescence imaging with conventional fluorescent probes. *Angew. Chem. Int. Ed. Engl.* **2008**, *47*, 6172–6176. [CrossRef]
11. Heintzman, R.; Ficz, G. Beaking the resolution limit in light microscopy. *Brief. Funct. Genom. Proteom.* **2006**, *5*, 289–301. [CrossRef]
12. Betzig, E.; Patterson, G.H.; Sougrat, R.; Lindwasser, O.W.; Olenych, S.; Bonifacino, J.S.; Davidson, M.W.; Lippincott-Schwartz, J.; Hess, H.F. Imaging intracellular fluorescent proteins at nanometer resolution. *Science* **2006**, *313*, 1642–1645. [CrossRef]
13. Hess, T.S.; Girirajan, P.K.T.; Mason, D.M. Ultra-high resolution imaging by fluorescence photoactivation localization microscopy. *Biophys. J.* **2006**, *91*, 4258–4272. [CrossRef]
14. Huang, B. Super-resolution optical microscopy: Multiple choices. *Curr. Opin. Chem. Biol.* **2010**, *14*, 10–14. [CrossRef] [PubMed]
15. Rust, M.J.; Bates, M.; Zhuang, X. Sub-diffraction-limit imaging by stochastic optical re-construction microscopy (STORM). *Nat. Methods* **2006**, *3*, 793–795. [CrossRef] [PubMed]
16. Maslov, A.V.; Astratov, V.N. Imaging of sub-wavelength structures radiating coherently near microspheres. *Appl. Phys. Lett.* **2016**, *108*, 051104. [CrossRef]
17. Kassamakov, I.; Lecler, S.; Nolvi, A.; Leong-Hoi, A.; Montgomery, P.; Haeggstrom, E. 3D Super-Resolution Optical Profiling Using Microsphere Enhanced Mirau Interferometry. *Sci. Rep.* **2017**, *7*, 3683. [CrossRef]
18. Pendy, J.B. Negative refraction makes a perfect lens. *Phy. Rev. Lett.* **2000**, *85*, 3966–3969. [CrossRef]
19. Sirohi, R.S. *Speckle Metrology*; Marcel Dekker: New York, NY, USA, 1993; pp. 99–234.
20. Cloud, G. *Optical Methods of Engineering Analysis*; Cambridge University Press: New York, NY, USA, 1995; pp. 395–476.
21. Malacara, D. *Optical Shop Testing*; John Wiley &Sons: New York, NY, USA, 1992; pp. 501–652.
22. Arai, Y. Three-dimensional shape measurement beyond the diffraction limit of lens. *J. Mod. Opt.* **2018**, *65*, 1866–1874. [CrossRef]
23. Arai, Y. Three-Dimensional shape measurement beyond diffraction limit for measurement of Dynamic Events. In *Progress in Optomechatronic Technologies, Springer Proceedings in Physics*; Springer: Singapore, 2019; p. 233. [CrossRef]
24. Arai, Y. Precise wide-range three-dimensional shape measurement method to measure superfine structures based on speckle interferometry. *Opt. Eng.* **2020**, *59*, 014108. [CrossRef]
25. Arai, Y. Consideration of existence of phase information of object shape in zeroth-order diffraction beam using electromagnetic simulation with aperture in front of object. *J. Mod. Opt.* **2020**, *67*, 523–530. [CrossRef]
26. Arai, Y. Factors affecting the measurement resolution of super-resolution techniques based on speckle interferometry. *J. Mod. Opt.* **2022**, *69*, 897–910. [CrossRef]

27. Kohler, H. On Abbe's theory of image formation in the microscope. *Opt. Acta* **1981**, *28*, 1691–1701. [CrossRef]
28. Zhang, Y.; Luo, J.; Li, J.; Mao, D.; Zhang, Y.; Huang, Y. Fast Inverse-Scattering Reconstruction for Airborne High-Squint Radar Imagery Based on Doppler Centroid Compensation. *IEEE Trans. Geosci. Remote Sens.* **2021**, *60*, 5205517. [CrossRef]

photonics

MDPI

Article

Ensquared Energy and Optical Centroid Efficiency in Optical Sensors: Part 2, Primary Aberrations

Marija Strojnik [1,*], Robert Martin [2,†] and Yaujen Wang [3,‡]

1 Optical Research, Leon de los Aldama 37150, Gto., Mexico
2 Independent Researcher, West Hills, CA 91304, USA
3 Independent Researcher, Arcadia, CA 91006, USA; lightw123@gmail.com
* Correspondence: mstrojnik@gmail.com; Tel.: +1-480-479-7817
† Portion of this work was performed when the author was affiliated with: Lockheed-Martin Missiles & Space Company, Sunnyvale, CA 94088, USA.
‡ Portion of this work was performed when the author was affiliated with: Northrop Grumman, Aerospace Systems, Azusa, CA 91702, USA.

Abstract: We previously proposed that the optical centroid efficiency (*OCE*) might be a preferred figure-of-merit to the enclosed energy of a rectangular pixel (*EOD*) for an instrument subject to unpredictable environmental jitter and alignment conditions. Here we follow the same symbols for the corresponding quantities, particularly the width of the pixel as being equal to 2d. Here we analyze the performance of the *OCE* vs. the *EOD* for the three Seidel primary aberrations of an optical component: spherical, coma, and astigmatism, plus defocus. We show that the *OCE* has an approximate *U*-shape when graphed against the *EOD*, for the aberrations ranging from 0 to 1.25λ. We conclude that for pixels larger than $2d = 3\lambda F/\#$, a small pixel will feature better performance when expecting jitter, misalignment, and other environmental and unpredictable conditions. When evaluating the performance of low-aberration instruments in dynamic and unpredictable environments, the choice of the lager pixel $2d = 7\lambda F/\#$ might be advantageous. Its selection will result in the deterioration of image resolution.

Keywords: optical centroid efficiency; *OCE*; ensquared energy (energy on detector; *OED*); sensors; opto-mechatronics; opto-mechanics; numerical simulations; optical devices; spherical aberration; coma; astigmatism; defocus; environmental conditions; vibration; jitter; misalignment

1. Introduction

In a modern sensor, the perturbations of figure-of-merit include inherent design aberrations, manufacturing uncertainties and misalignment, and some unrealized environment impacts, including jitter, and other unpredictable opto-mechatronics conditions. In the previous work, the correlation between the detector pixel sizes as well as the central obscuration of the optical sensor and the *OCE* has been presented in detail [1]. In order to explore more characteristics of the *OCE*, in this follow-up study of previous work some of the most prominent aberrations in a sensor are included. Please note that when dealing with pixel size, we do not study sampling, which is a different phenomenon [2,3]. Similarly, we do not consider any image processing and enhancement that might facilitate or improve the acquisition of the information content in the image [4–7].

2. Aberration in an Optical Sensor

Descriptions of optical aberrations may be founded in many optical textbooks [8–11]. In general, the five Seidel sum, "spherical, coma, astigmatism, Petzval curvature, distortion and defocus", are an excellent representation. In this work we focus on the evaluation of *OCE* caused by each of the three typical aberrations (spherical, coma, and astigmatism) and defocus in Sections 3–6. The image is located at the best focus. Also, all the terminologies employed in the following sections are identical to the previous work.

We investigate the relationship between the *OCE* and the *EOD*. The *prf* is based on the chief ray selection. The centroid is calculated after statistical averaging. All aberration studies are performed for two different pixel sizes, $2d = 3\lambda F/\#$ and $2d = 7\lambda F/\#$, where the linear detector pixel dimension is 2d. This choice corresponds to a relatively small and relatively large pixel size in comparison to the diameter of the Airy disc. The former is a bit larger than the Airy disc diameter, while the latter is somewhat smaller than three Airy discs. In *Code V*, the aberration represented with an appropriate coefficient in Zernike polynomials for each aberration is added at the entrance pupil of our optical model.

3. Spherical Aberration Studies, $a_{40}\rho^4$

3.1. Spherical for Small Pixel Values, $2d = 3\lambda F/\#$

We investigate the relationship between the *OCE* and the *EOD* for the increase in spherical aberration from $a_{40} = 0$ to $a_{40} = 1.25\lambda$. The results for small pixel size $2d = 3\lambda F/\#$ are presented as the *OCE* vs. *EOD* graph in Figure 1. The *EOD* values are arranged in the ascending order from about *0.3* to about *0.9*, corresponding to the decrease in the amount of the spherical aberration from bottom left ($a_{40} = 1.25\lambda$) to top right ($a_{40} = 0$) of the graph. As the amount of spherical aberration increases, the energy in the central peak of the *psf*-s is pushed out symmetrically to the side lobes in the radial direction. The addition of spherical aberration results in the reduction of the *EOD*.

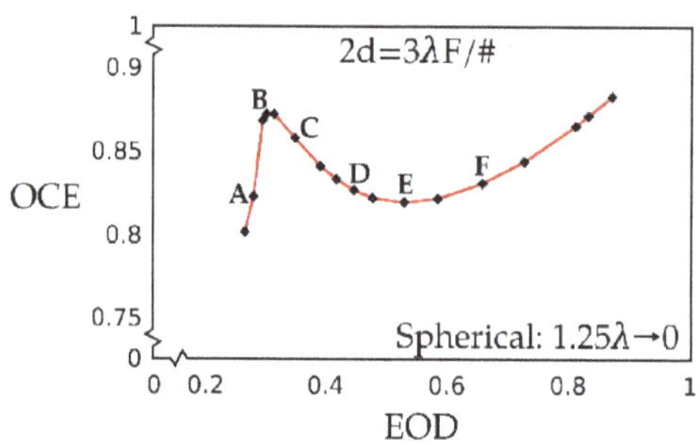

Figure 1. *OCE* vs. *EOD* for a lens with the spherical aberration (ρ^4-term) for a detector pixel size of 2d = $3\lambda F/\#$. The amount of spherical aberration decreases from the lower left corner with the value of $a_{40} = 1.25\lambda$ along the curve to the upper right corner with $a_{40} = 0$.

When there is no aberration, the *OCE* achieves a value of 0.88 for the maximum energy on the detector enclosed by the square pixel, and an *EOD* value of 0.88 for pixel size $2d = 3\lambda F/\#$. The *OCE* first smoothly decreases with increasing amount of aberration, arriving at a local minimum of 0.82 for the *EOD* of 0.58. Then *OCE* climbs to a sharp peak, nearly equal to the case of zero (0) aberration around the aberration value $a_{40} = 1\lambda$. When the aberration further increases, the lowest value is achieved for the *OCE*, equal to 0.80 for an *EOD* of 0.25.

Here we note that the *OCE* is proportional to the *EOD* for about half of the *EOD* interval and inversely proportional over the other half. We may simply observe that they are not correlated. We use the description that two quantities are correlated when they either increase or decrease under similar circumstances. Two quantities are not correlated when their changes occur under different circumstances.

For a small pixel size of $2d = 3\lambda F/\#$, the *OCE* first rapidly increases with increasing values of *EOD*, reaching a maximum of 0.872 when the *EOD* is equal to 0.31. The low value of the *EOD* is caused by the fact that the energy moves from the central spot to the higher-order rings for a high amount of aspherical aberration. When the optical axis relative to the pixel position moves around in the *OCE* determination, the position of the image centroid has no deteriorating consequence because the amount of energy is about the same all over the pixel surface. The *OCE* value is relatively high because the normalization with the *EOD* is included in its definition.

The *OCE* peak around 1λ aberration is followed by a sharp decrease in the *OCE* values when the *EOD* is further increased, until a broad minimum is attained at *EOD* = *0.53*, with the still-high value of 0.82 for the *OCE*. In this region, the amount of spherical aberration has decreased, so the spot is becoming better defined, resulting in increased energy collection over the pixel surface. A more compact spot results in smaller average energy on detector for a displaced pixel center and increased energy on detector for the centered image. Both effects combine to decrease the *OCE* for small pixel size $2d = 3\lambda F/\#$. Here, the displacement of the spot position over the pixel area is very sensitive. Only with decreasing aberration values are the *OCE* and the *EOD* correlated as expected when the spot over the pixel surface becomes compact under conditions of no or a small amount of aberration.

Figure 2 presents cross-sections of the instrument point response function *prf* as a function of a radial coordinate, normalized to pixel size for the points labeled in Figure 1, that is, *A*, *B*, *C*, *D*, *E*, and *F*.

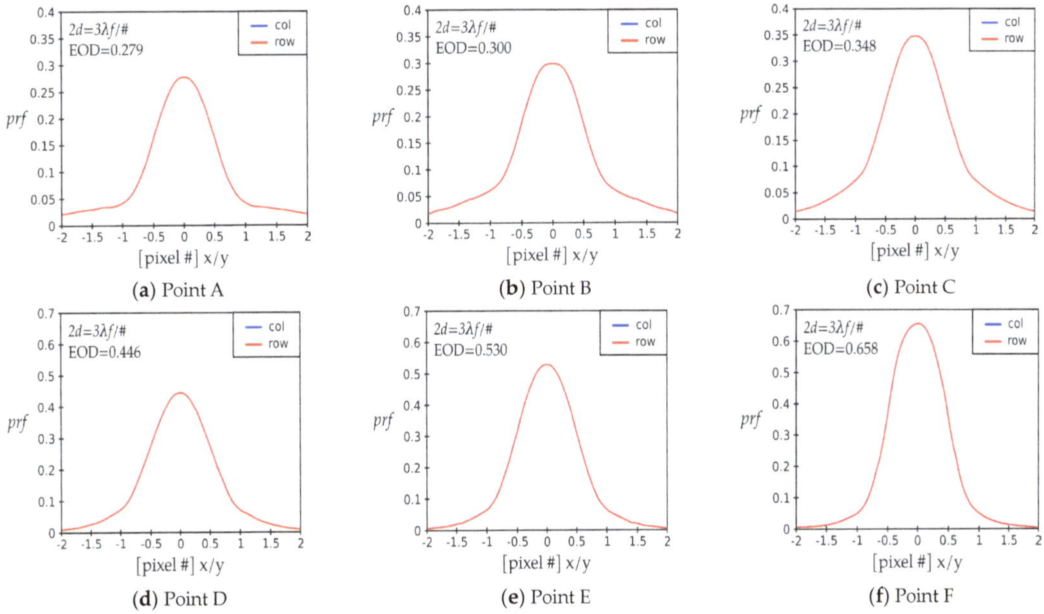

Figure 2. The instrument point response function *prf* corresponding to selected points in Figure 1 as a function of two orthogonal directions, x and y, or rows and columns, for the case of spherical aberration. The amount of aberration decreases from about $a_{40} = 1\lambda$ to $a_{40} = 0.5\lambda$, going from (**a**–**f**). With the decrease in the aberration, the power in the image becomes increasingly more centralized. The figures are arranged in order of increasing energy on the detector, which corresponds to a decreasing amount of aberration. The graphs in the first row correspond to the peak in the *OCE* vs. *EOD* graph in Figure 1, while those in the second row correspond to the valley. The common features of the *prf*-graphs for the peak are a decreased peak value, an increased amount of the compact support, and the fact that the images carry a significant amount of aberration. The valley is characterized by a relatively high peak value, the absence of compact support, and a decreased amount of aberration. Due to the symmetry of the problem, the red line overlaps over the blue line.

First, we study the three points near the narrow and sharp peak, that is, Points *A*, *B*, and *C*. The corresponding *prf*-graphs are plotted in Figure 2a–c. At first glance, these three graphs, lined in the first row, appear to be quite like the ones in the second row.

The most important difference to note is the relatively large difference in the *prf*-values between the center (0) and edge of the detector pixel (0.5). Point *B* has the highest difference between these two values; therefore, it is located near the local peak for the lower *EOD* regime. The *prf*-value at the pixel center (0.) is equal to *0.30*. The value of the *prf* at the pixel edge (*0.5*) is *0.17*. The difference in the *prf*-values at the center and the edge of the detector pixel is *0.13*, for a small pixel value of $2d = 3\lambda F/\#$.

The point *A* is located at nearly the lowest value of *the EOD*, *0.279*. The *prf*-value at the pixel center is equal to *0.279*. The value of the *prf* at the pixel edge (*0.5*) is *0.22*. The difference in the *prf*-values at the center and the edge of the detector pixel is *0.123*, a value smaller than that at Point *B;* therefore, the *OCE* is also smaller for Point *A* than for Point *B*. Point *C* is located at the value of the *EOD* higher than that for the highest point (*B*), at *0.345*. The *prf*-value at the pixel center is equal to *0.345*. The value of the *prf* at the pixel edge (*0.5*) is *0.24*. The difference in the *prf*-values at the center and the edge of the detector pixel is *0.105*, a value smaller than that of Point *B*; therefore, the *OCE* is also smaller for Point *C* than for Point *B*.

Next, we investigate the *OCE* vs. *EOD* relationship around the minimum. Figure 2d–f display the three *prf* plots for Points *D*, *E*, and *F* around the minimum. Point *F*, featured on Figure 2f, exhibits a peak of the *prf* graph, that is, the *EOD*, of *0.66*. We see that the value of the *prf* at the pixel edge (*0.5*) is 0.36. The difference in the *prf* at the center, or the *EOD*, and that the edge of the pixel (*0.5*) is *0.30*.

We examine the minimum next, approaching it from the large EOD values. Point *E*, at the very minimum of the *OCE* vs. *EOD* curve is presented in Figure 2e. It features an *EOD* value of *0.53*, a value smaller than that of Point *F*. We see that the value of the *prf* at the pixel edge (*0.5*) is *0.33*. This *prf* graph exhibits a relatively small difference in the *prf* values between the center and edge of the pixel of *0.20*. Therefore, the value of the *OCE* is the lowest there. Figure 2d features the *prf* for point *D*. The peak of the *prf*, the *EOD*, is low among the triplets, at *0.446*. The value of the *prf* at the pixel edge (*0.5*) is *0.32*. This is a relatively smaller difference in the *prf*-values between the center and edge of the pixel of 0.13. Because of the relatively low peak value of the prf at the center and the relatively high value of the *prf* at the pixel edge, Point *D* has a higher *OCE* value than the neighboring Point *E*.

3.2. *Spherical with Large Pixel Values, $2d = 7\lambda F/\#$*

The next interesting result is observed when the detector size is increased from $2d = 3\lambda F/\#$ to $2d = 7\lambda F/\#$. Figure 3 presents the *OCE* vs. the *EOD* for a lens with the spherical aberration (a_{40} -- term) for the pixel size $2d = 7\lambda F/\#$. For the larger pixel size discussed here, we start graphing the *OCE* with the value of *0.4* for the *EOD* with spherical aberration equal to $a_{40} = 1.25\lambda$. Traveling along the curve, the spherical aberration decreases in the upper right corner to zero.

General features of the *OCE* vs. *EOD* curve in Figure 3 for large pixel $2d = 7\lambda F/\#$ are quite similar to those for the small pixel in Figure 1. The *OCE* values increase for very small values of the *EOD* and for large values of the *EOD* with the increasing amount of aberration. A peak and a valley arise between these two regions, like the shape in Figure 1. The amount of spherical aberration decreases from the lower left corner with the value of $a_{40} = 1.25\lambda$ along the curve to the upper right corner with $a_{40} = 0$.

In the absence of aberration, the maximum energy on the detector enclosed by the square pixel is *0.95*. When the amount of spherical aberration increases from 0 to $a_{40} = 1.25\lambda$, the *EOD* decreases from about *0.95* to *0.4* when using a pixel size of $2d = 7\lambda F/\#$. When there is no aberration, the *OCE* achieves the value of *0.89*. The *OCE* first rapidly decreases with increasing amount of aberration, achieving the valley bottom of the *OCE* = *0.78* at the *EOD* = *0.80*. The *OCE* then, with a slope of about (minus) 45 degrees, approaches and forms a rounded peak, nearly equal to the case of no aberration around the aberration value of $a_{40} = 1\lambda$. When the aberration further increases, about an average value is achieved for the *OCE*, equal to *0.84* for the *EOD* of *0.4*. Here we note that the *OCE* is proportional to the *EOD* in about one half of the *EOD* interval and inversely proportional in the other half. They are not correlated.

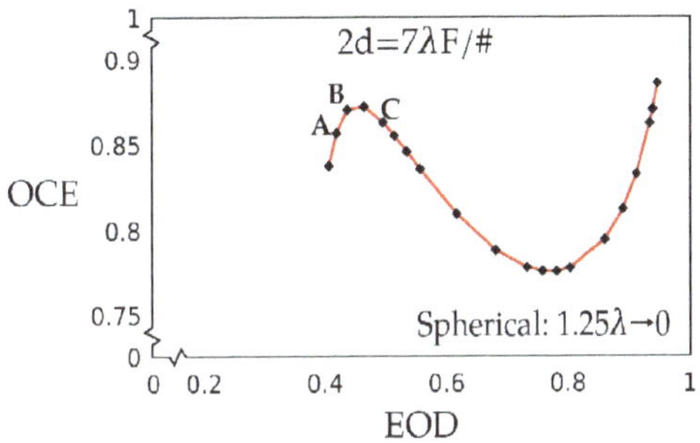

Figure 3. *OCE* vs. *EOD* for a lens with the spherical aberration (ρ^4-term) for a detector pixel size of $2d = 7\lambda F/\#$. The amount of spherical aberration decreases from the middle-left region with the value of $a_{40} = 1.25\lambda$ along the curve to the upper right corner with $a_{40} = 0$.

Next, we endeavor to explain the first one-third of the curve for the small *EOD* values, featuring a narrow, sharp *OCE* peak. The *prf*-s corresponding to the three points around the local peak (Points *A*, *B*, and *C*) are plotted in Figure 4. Point response functions *prf*-s for Points *A*, *B*, and *C* have a similar profile across the pixel surface. For these points, the *EOD* increases from *0.42* through *0.47* to *0.50*. The *prf(0)* and *prf(0.5)* for Point *A* are *0.421* and *0.25*, resulting in their difference of *0.17*. The *prf(0)* and *prf(0.5)* for Point *B* are *0.47* and 0.31, resulting in their difference of 0.16. Even so, Point *A* has a lower *EOD* value than Point *B*. This leads to a relative higher variation of the *EOD* across the pixel surface and therefore a lower *OCE* value for Point *A*.

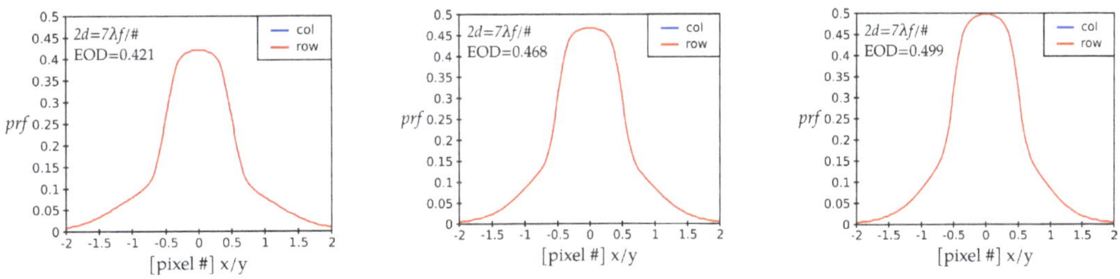

(a) Point A **(b)** Point B **(c)** Point C

Figure 4. The instrument point response function *prf* corresponding to selected points in Figure 3 as a function of two orthogonal directions, x and y, or rows and columns, for the case of spherical aberration. We model a case where the detector pixel size equals $7\lambda F/\#$. With an increase in the aberration, the energy spreads out. The energy on detector *EOD* for the same pixel size decreases with increasing amount of aberration. The figures are arranged in order of increasing energy on the detector, which corresponds to a decreasing amount of aberration. Due to the symmetry of the problem, the red line overlaps over the blue line.

The *prf*-curve corresponding to Point C has the highest *EOD* value (that is, *prf(0)*), and a more narrowly formed cone at the center than the other two. Additionally, it also features a relatively larger difference in the *prf*-values between the center and edge of the detector pixel than that of Point *B*. The edge value is *0.32*, and the center value is 0.5, resulting in the difference from the center to the edge of *0.18*. Thus, Point *C* presents a larger variation in the *prf*-values across the pixel surface. Therefore, Point *C* ends up having a somewhat lower *OCE* value than Point *B*.

In the presence of 3rd order spherical aberration in an otherwise ideal lens-optical detector combination, the *EOD* decreases in the presence of aberration because the energy in the central spot is being pushed out to the outer rings. The amount of energy that is incident on a given pixel and the average *prf* across the detector determine the relationship between *EOD* and *OCE*. For a different size pixel, with its value somewhere between $2d = 3\lambda F/\#$ and $2d = 7\lambda F/\#$, the initial points and the end points would be lying between the corresponding points on Figures 1 and 3. The *OCE*-vs-*EOD* curve would exhibit a shape similar to the two limiting cases presented here and lying in-between them.

4. Coma-y: $a_{31}\ \rho^3 sin\theta$

4.1. Coma-y for Small Pixel Size, $2d = 3\lambda F/\#$

We present the *OCE* vs. the *EOD* graph for the case of coma aberration for small pixel size $2d = 3\lambda F/\#$ in Figure 5. In the absence of aberration, the maximum energy on the detector enclosed by the square pixel is 0.88. When the amount of coma aberration increases from $a_{31} = 0$ to $a_{31} = 1.25\lambda$, the *EOD* decreases from about 0.88 to 0.25 when using a pixel size of $2d = 3\lambda F/\#$. Except for the outer regions of the interval under study, the increase in the amount of aberration appears to affect the EOD moderately.

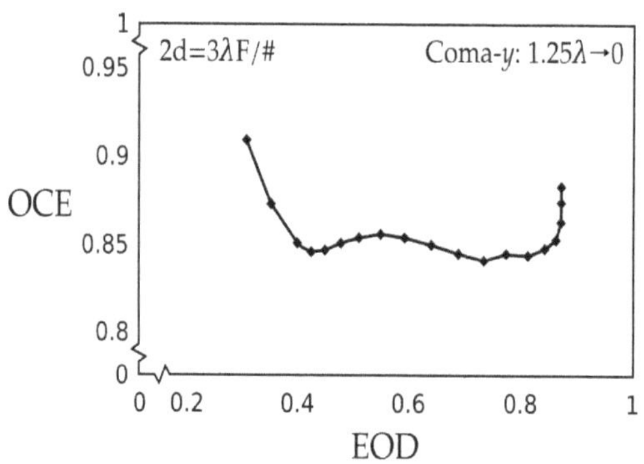

Figure 5. *OCE* vs. *EOD* for a lens with the coma aberration ($a_{31} = \rho^3 sin\theta$) for a detector pixel size of $2d = 3\lambda F/\#$. The amount of coma aberration decreases from the upper left corner with the value of $a_{31} = 1.25\lambda$ along the curve to the right middle area with $a_{31} = 0$.

When there is no aberration, the *OCE* achieves the value of 0.88. At first the *OCE* sharply, nearly vertically, decreases with increasing amount of aberration; then, it rolls off to an approximately constant value, at about $OCT = 0.85$. It features two small valleys of $OCT = 0.84$, at $EOD = 0.74$ and 0.41. The *OCE* starts to increase again with increasing aberration at about $a_{31} = 1\lambda$. When the aberration further increases to $a_{31} = 1.25\lambda$, the *OCE* increases, achieving a value of 0.91 for an *EOD* of 0.25.

Here we note that the *OCE* is roughly independent of the *EOD* for the middle half of the *EOD* interval. During the first quarter of the interval, it is inversely proportional. It is proportional in the last quarter. We can also say that the *OCE* and the *EOD* are independent of each other in the case of the coma aberration. We may simply observe that coma affects differently the *EOD* and the *OCE*. We may simply observe that the *OCE* and the *EOD* are not correlated under the effects of coma for small pixel size in the study.

4.2. Coma-y with Large Pixel Value, $2d = 7\lambda F/\#$

We present the *OCE* vs. the *EOD* graph for the case of coma aberration for large pixel size $2d = 7\lambda F/\#$ in Figure 6. In the absence of aberration, the maximum energy on the detector enclosed by the square pixel is *0.95*. When the amount of coma aberration increases from $a_{31} = 0$ to $a_{31} = 1.25\lambda$, the *EOD* decreases from about *0.96* to *0.6* when using a pixel size of $2d = 7\lambda F/\#$, while the *OCE* decreases from *0.95* to *0.83*.

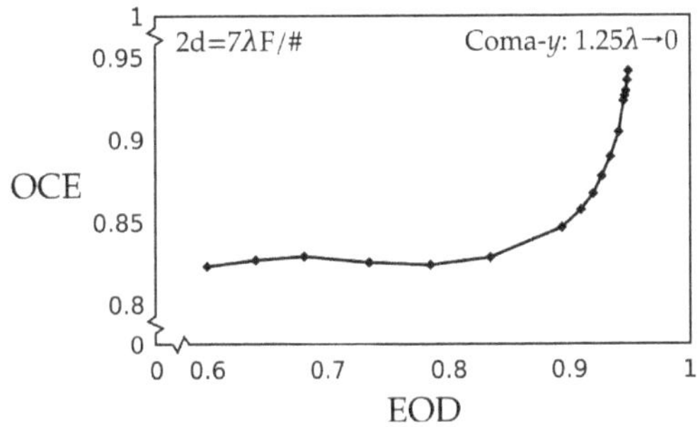

Figure 6. *OCE* vs. *EOD* graph for a lens with the coma aberration ($a_{31}\ \rho^3 \sin\theta$) for a detector pixel size of 2d = $7\lambda F/\#$. The amount of coma aberration decreases from the lower left corner with the value of $a_{31} = 1.25\lambda$ along the curve to the upper right corner with $a_{31} = 0$.

In absence of aberration, the *OCE* achieves the value of *0.95*. From there, it first rapidly decreases with increasing amount of aberration, arriving to its elbow at the *EOD = 0.88*. The *OCE* then remains approximately constant with increasing aberration, achieving a small hump for the aberration value of $a_{31} = 1\lambda$ for the *EOD* of *0.68*. When the aberration further increases, the *OCE* decreases to *0.825* for the *EOD* value of *0.6*.

Here we note that the *OCE* is roughly independent of the *EOD* for more than one half of the *EOD* interval. During about the last third of the interval, the *OCE* is proportional to the *EOD*. We may simply conclude that for large amounts of coma aberration, the *OCE* and the *EOD* are independent of each other. For small amounts of the coma aberration, the *OCE* is proportional to the *EOD*.

5. Astigmatism: $a_{22}\ \rho^2 cos^2\theta$

5.1. Astigmatism $a_{22}\ \rho^2 cos^2\theta$ for Small Pixel Size, $2d = 3\lambda F/\#$

We present the *OCE* vs. the *EOD* graph for the case of astigmatism aberration for small pixel size $2d = 3\lambda F/\#$ in Figure 7.

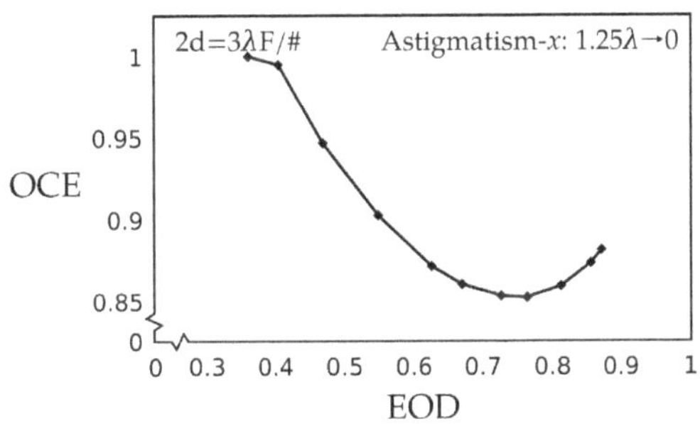

Figure 7. *OCE* vs. *EOD* graph for a lens with the astigmatism aberration ($a_{22}\ \rho^2 cos^2\theta$) for a detector pixel size of $2d = 3\lambda F/\#$. The first zero of the Bessel function is located at $1.22\ \lambda F/\#$. In the absence of aberration, the pixel encloses somewhat more than the Airy disc diameter. The amount of astigmatism aberration decreases from the upper left corner with the value of $a_{22} = 1.25\ \lambda$ along the curve to the right middle area with $a_{22} = 0$.

In the absence of aberration, the maximum energy on the detector enclosed by the square pixel is *0.88*. When the amount of astigmatism aberration increases from $a_{22} = 0$

to a_{22} = *1.25* λ, the *EOD* decreases from about *0.88* to *0.35* when using a pixel size of *2d* = *3*λ*F/#*. When there is no aberration, the *OCE* achieves the value of *0.88*. For an initial short interval, the *OCE* first smoothly decreases with increasing amount of astigmatism aberration, achieving the minimum value of *0.85* for the *EOD* = 0.74; afterwards, it climbs with about a 45-degree slope to its rounded peak at practically *1*, while the *EOD* is only *0.35* for the aberration value a_{22} = *1.25*λ.

For about the first two-thirds of the *EOD* interval, the *OCE* is inversely proportional to the *EOD*. During the last third, the *OCE* is proportional to the *EOD*. We note that the general curve has the same shape as the central part of the corresponding spherical aberration curve. We may simply observe that the *OCE* and the *EOD* are not correlated for small pixel sizes.

5.2. Astigmatism $a_{22}\ \rho^2 cos^2\theta$ for Large Pixel Value, 2d = 7λF/#

We present the *OCE* vs. the *EOD* graph for the case of astigmatism aberration for large pixel size *2d* = *7*λ*F/#* in Figure 8.

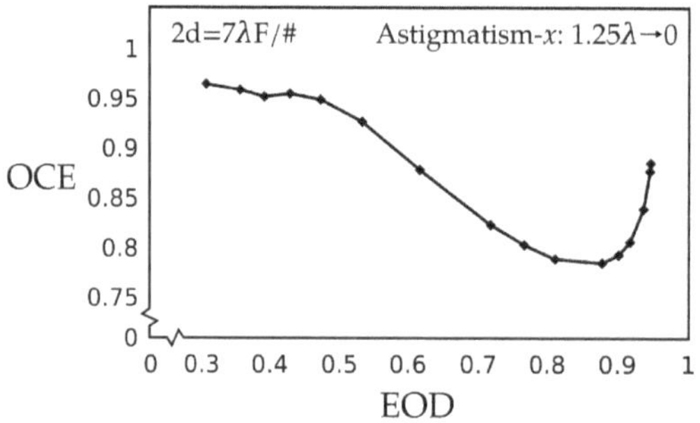

Figure 8. *OCE* vs. *EOD* graph for a lens with the astigmatism aberration ($a_{22}\ \rho^2 cos^2\theta$) for a detector pixel size of *2d* =*7*λ*F/#*. The first zero of the Bessel function is located at *1.22* λ*F/#*. In the absence of aberration, the pixel encloses *2.87* Airy discs (nearly 3). The amount of astigmatism aberration decreases from the upper left corner with the value of a_{22} = *1.25*λ along the curve to the right middle area with a_{22} = *0*.

In the absence of aberration, the maximum energy on the detector enclosed by the square pixel is *0.95*. When the amount of astigmatism aberration increases from a_{22} = 0 to a_{22} = *1.25*λ, the *EOD* decreases from about 0.95 to 0.35 when using a pixel size of *2d* = *7*λ*F/#*. When there is no aberration, the *OCE* achieves the value of *0.88*. For an initial short interval, the *OCE* first smoothly decreases with increasing amount of astigmatism aberration, achieving the minimum value of *OCE* = *0.78* at the *EOD* = *0.86*; afterwards, it climbs with about a 45-degree slope to its elongated rounded peak for the *OCE* = *0.97* at the *EOD* = *0.3*, for the aberration value a_{22} = *1.25*λ.

For about the fourth or fifth *EOD* interval, the *OCE* is inversely proportional to the *EOD*. During the last fifth, the *OCE* is proportional to the *EOD*. We note that the general curve has a distorted shape of the central part of the corresponding spherical aberration curve. We may observe that the *OCE* and the *EOD* are not correlated for large pixels.

6. Defocus/Power: $a_{20}\ \rho^2$

6.1. Defocus/Power $a_{20}\ \rho^2$ for Small Pixel 2a = 3λF/#

We present the *OCE* vs. the *EOD* graph for the case of defocus aberration for small pixel size *3*λ*F/#* in Figure 9. In the absence of defocus aberration, the maximum energy on the detector enclosed by the square pixel is 0.88. When the amount of defocus aberration increases from a_{20} = *0* to a_{20} = *1.25*λ, the *EOD* decreases from about *0.88* to *0.31* when using a pixel size of *2d* = *3*λ*F/#*. When there is no aberration, the *OCE* achieves the value of

0.88. The *OCE* first smoothly decreases with increasing amount of aberration, achieving a minimum of *0.82* for the *EOD* of 0.62. From there, the curve looks like a quadratic function tilted to the side.

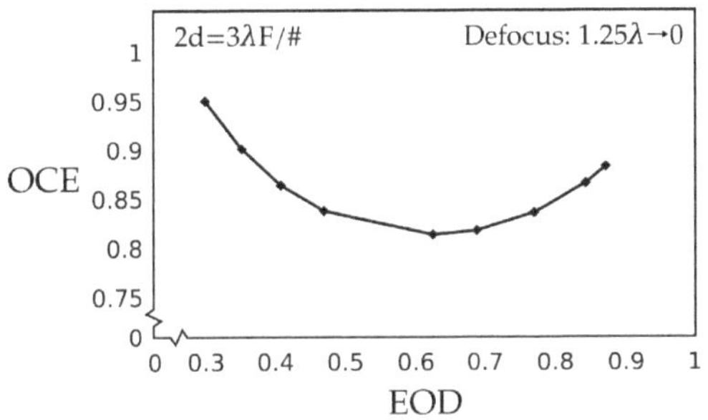

Figure 9. *OCE* vs. *EOD* graph for a lens with the defocus aberration ($a_{20}\,\rho^2$) for a detector pixel size of $2d = 3\lambda F/\#$. The amount of defocus aberration decreases from the upper left corner with the value of $a_{20} = 1.25\,\lambda$ along the curve to the right middle region with $a_{20} = 0$.

When the aberration further increases, the highest value is achieved for the *OCE* of *0.95* at the *EOD* of *0.31*. Here we note that for defocus aberration, the *OCE* is proportional to the *EOD* for half of the *EOD* interval and inversely proportional to the other half. We may simply observe that they are not correlated.

6.2. Defocus/Power with Large Pixel Value, 2d = 7λF/#

We present the *OCE* vs. the *EOD* graph for the case of defocus aberration for large pixel size $2d = 7\lambda F/\#$ in Figure 10.

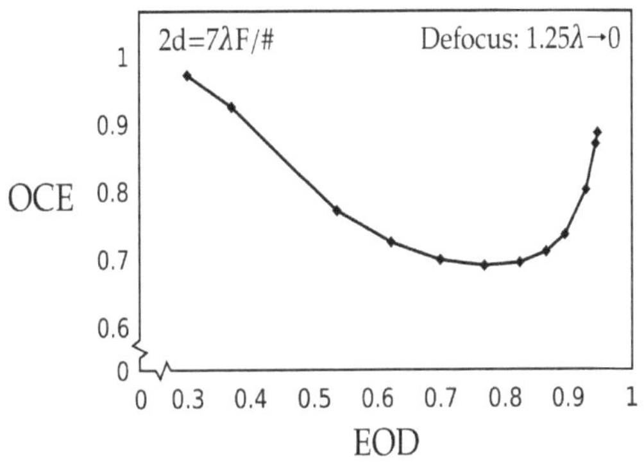

Figure 10. *OCE* vs. *EOD* graph for a lens with the defocus aberration (a_{20}) for a detector pixel size of $7\lambda F/\#$. The amount of defocus aberration decreases along the curve from the upper left corner through the minimum and then to the upper right corner with $a_{20} = 0$.

In the absence of defocus aberration, the maximum energy on the detector enclosed by the square pixel is *0.95*. When the amount of defocus aberration increases from $a_{20} = 0$ to $a_{20} = 1.25\lambda$, the *EOD* decreases from about *0.95* to *0.3* when using a pixel size of $2d = 7\lambda F/\#$. In the absence of aberration, the *OCE* achieves the value of *0.89*. It first rapidly decreases with increasing amount of aberration, achieving a minimum of *0.68* at the *EOD* of *0.77*. From there, the curve looks like a quadratic function with unequal sides. The *OCE* graph vs. *EOD* graph increases with about a 45-degree inclination angle with increasing aberration. When the aberration further increases, the highest value is achieved for the *OCE*, equal to 0.98 for the *EOD* of *0.31*.

Here we note that for the defocus aberration, the *OCE* is proportional to the *EOD* for a quarter of the *EOD* interval. It is inversely proportional for about three-quarters of the *EOD* interval. We may simply observe that the *OCE* and the *EOD* are not correlated for the defocus in the case of the large pixel.

7. Discussion

7.1. Spherical Aberration

We present the *OCE* vs. the *EOD* graph for the case of spherical aberration for two pixel sizes $2d = 3\lambda F/\#$ and $2d = 7\lambda F/\#$ in Figure 11. The first point on the left of each graph corresponds to the amount of spherical aberration, $a_{40} = 1.25\lambda$, while the last point on the right of the graph corresponds to the case of no aberration, $a_{40} = 0$.

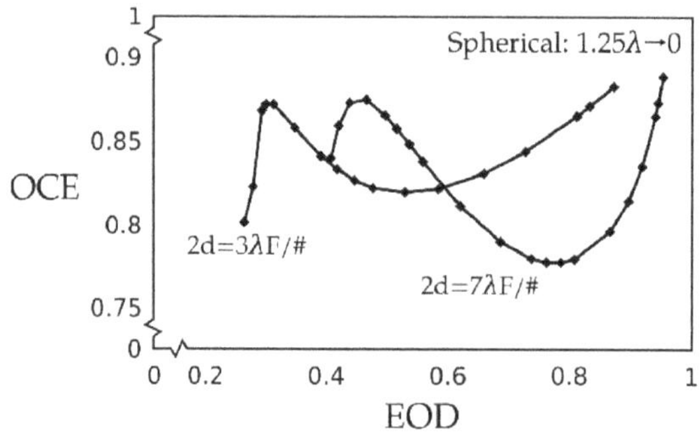

Figure 11. *OCE* vs. *EOD* graph for the case of spherical aberration for two pixel sizes $2d = 3\lambda F/\#$ and $2d = 7\lambda F/\#$. The first point on the left of each graph corresponds to the amount of spherical aberration, $a_{40} = 1.25\ \lambda$, while the last point on the right of each graph corresponds to the case of no aberration, $a_{40} = 0$.

The most interesting aspect in comparison of these two graphs is that they are quite similar in general shape. The graph for the small pixel appears to be compressed along the *OCE* axis and displaced toward smaller values along the *EOD* axis. They both feature peaks, corresponding to the absence of aberration that has nearly the same value, $OCE = 0.87$. The peak is displaced for the smaller pixel by about $DEOD = 0.12$. Both graphs also include valleys, displaced by $DEOD = 0.2$ and $DOCE = 0.12$. The minimum for the larger pixel is deeper and displaced to larger *EOD* values.

In the absence of aberration, the maximum energy on the detector enclosed by the square pixel is *0.88* for a small pixel size of $2d = 3\lambda F/\#$. The maximum energy on the detector for the larger pixel size of $2d = 7\lambda F/\#$ increases to *0.95*. This increment of *0.07* is reasonable because a larger pixel encloses more energy for an aligned system. In the absence of aberration, the pixel size has minimal effects on possible displacement of the optical axis on the collected energy or other environmental effects. From the local maximum characterized by no aberration, graphs then decrease to a valley, with the graph for the large pixel size having a slope larger than 45 degrees and that for the smaller pixel having a smaller slope.

When the amount of spherical aberration increases from $a_{40} = 0$ *to* $a_{40} = 1.25\lambda$, the *EOD* decreases to *0.25* for a pixel size of $2d = 3\lambda F/\#$. At the same time, the *EOD* increases to 0.4 for pixel size $2d = 7\lambda F/\#$. This increase of *0.15* for the larger pixel in the enclosed energy is also reasonable because a large amount if spherical aberration spreads the energy to a larger spot. A small pixel intercepts a smaller fraction of energy of a large spot than a large pixel; therefore, the small pixel is impacted by aberration more than the large one. The increase in the *EOD* for the large pixel is twice the increase for the small pixel. Because the displacement of the optical axis still allows a large amount of energy to be intercepted.

7.2. Coma-y

We present the curve of the *OCE* vs. the *EOD* graph for the case of coma-y aberration for two pixel sizes ($2d = 3\lambda F/\#$ and $2d = 7\lambda F/\#$) in Figure 12. The first point on the left of each graph corresponds to the amount of coma, $a_{31} = 1.25\ \lambda$, while the last point on the right of the graph corresponds to the case of no aberration, $a_{31} = 0$.

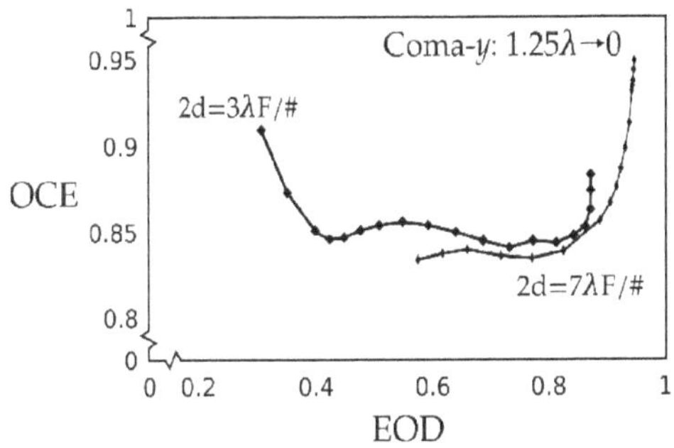

Figure 12. *OCE* vs. *EOD* graph for the case of coma-y aberration for two pixel sizes $2d = 3\lambda F/\#$ and $2d = 7\lambda F/\#$. The first point on the left of each graph corresponds to the amount of coma aberration, $a_{31} = 1.25\lambda$, while the last point on the graph corresponds to the case of no aberration, $a_{31} = 0$. For a large portion of the aberration under study, from about $a_{31} = 0.25\lambda$ to roughly $a_{31} = 1.25\lambda$, the *OCE* seems to be independent of the amount of aberration.

In the absence of aberration, the *EOD* enclosed by the square pixel is *0.88*, for the small pixel size $2d = 3\lambda F/\#$. In the absence of aberration, the maximum energy on the detector enclosed by the square pixel is *0.95* for the case of the large pixel size, $2d = 7\lambda F/\#$. This is a small shift of *0.07*. When the amount of coma aberration increases to $a_{31} = 1.25\lambda$, the *EOD* decreases from to *0.25* in the case of a small pixel size $2d = 3\lambda F/\#$. When the amount of coma aberration increases from $a_{31} = 0$ to $a_{31} = 1.25\lambda$, the *EOD* decreases to *0.6* when using a pixel size of $2d = 7\lambda F/\#$. When there is no aberration, the *OCE* achieves the value of *0.98* for the small pixel size $2d = 3\lambda F/\#$. When there is no aberration, the *OCE* achieves a value of *0.95* in the case of the large pixel size $2d = 7\lambda F/\#$.

For the small pixel size $2d = 3\lambda F$, the *OCE* initially sharply vertically decreases with increasing amounts of aberration, taking a sharp turn for the *EOD* value of *0.84*. The *OCE* then rolls off to an approximately constant value, at about *0.85*, with two small valleys of *0.84* at *0.77* and *0.41*. F for the case of the large pixel size $2d = 7\lambda F/\#$, the *OCT* initially nearly vertically rapidly decreases with increasing amount of aberration, taking a sharp turn at the *EOD* value of *0.84*. The *OCE* then remains approximately constant, achieving a small hump for aberration value $a_{31} = 1\lambda$ for the *EOD* of *0.68*.

The *OCE* starts to increase again at the aberration of $a_{31} = 1\lambda$. When the aberration further increases, the *OCE* increases, achieving a value of *0.91* for an *EOD* of *0.25* for the small pixel size $2d = 3\lambda F/\#$. When the aberration further increases, the *OCE* decreases to *0.825* for the *EOD* value of *0.60*, for the case of the large pixel size $2d = 7\lambda F/\#$.

For a large portion of the aberration under study, from about $a_{31} = 0.25\lambda$ to roughly $a_{31} = 1.25\lambda$, the *OCE* seems to be independent of the amount of aberration. The *EOD* for the aberration $a_{31} = 1.25\lambda$ nearly doubles from small (*EOD = 0.32*) to the large pixel (*EOD = 0.60*). This means that the coma is very sensitive to the pixel size for large aberrations. For a small amount of aberration, the *EOD* has low sensitivity on the coma aberration, while the *OCE* changes rapidly with the aberration.

7.3. Astigmatism

We present the curve of the *OCE* vs. the *EOD* graph for the case of astigmatism aberration for two pixel sizes $2d = 3\lambda F/\#$ and $2d = 7\lambda F/\#$ in Figure 13. The first point on the left on each graph corresponds to the amount of astigmatism aberration $a_{22} = 1.25\lambda$, while the last point on the right side of the graph corresponds to the case of no aberration.

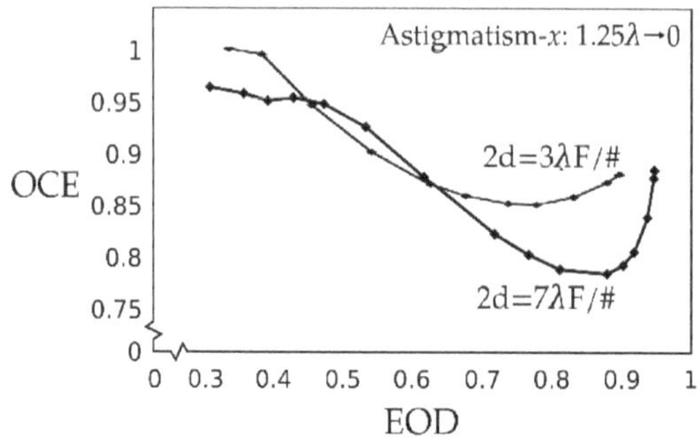

Figure 13. *OCE* vs. *EOD* graph in the case of astigmatism aberration for two pixel sizes, $2d = 3\lambda F/\#$ and $2d = 7\lambda F/\#$. The first point on the left of each graph corresponds to the amount of astigmatism aberration, $a_{22} = 1.25\lambda$, while the last point on the right of the graph corresponds to the case of no aberration, $a_{22} = 0$.

In the absence of aberration, the maximum energy on the detector enclosed by the square pixel is *0.88* for the small pixel size $2d = 3\lambda F/\#$. In the absence of aberration, the maximum *EOD* enclosed by the square pixel is *0.89*, for the case of the large pixel size $2d = 7\lambda F/\#$. When the amount of astigmatism aberration increases from *0* to *1.25* λ, the *EOD* decreases from about *0.88* to *0.35* when using a pixel size of $2d = 3\lambda F/\#$.

In the absence of aberration, the *OCE* achieves the value of 0.89 at the *EOD* of *0.88* for the small pixel size $2d = 3\lambda F/\#$. When there is no aberration, the *OCE* achieves the value of 0.89 for the large pixel size $2d = 7\lambda F/\#$. For the small pixel size $2d = 3\lambda F/\#$, the *OCE* first slowly and smoothly decreases with increasing amount of aberration, achieving the minimum value of 0.85 at about 0.74. The *OCE* first rapidly decreases with increasing amount of aberration, achieving the minimum at *0.78* at *EOD* of *0.84*, for the case of the large pixel size 2d = $7\lambda F/\#$.

After the valley, the *OCE*s climb with an about 45-degree slope to rounded peaks at different heights. Practically 1 is achieved for the case of the small pixel size $2d = 3\lambda F/\#$, at the *EOD* of only *0.35* for the aberration value *1.25* λ. In the case of the large pixel size $2d = 7\lambda F/\#$, the slope is somewhat higher, but the curve rounds off sooner to a nearly horizontal extended hump around 1λ. When the aberration further increases, the *OCE* slowly increases to a value of *0.96* for the *EOD* of *0.31*, for the case of the large pixel size $2d = 7\lambda F/\#$.

The shapes of two astigmatism curves for the small and large pixel are, in general, quite similar. For the small amount of aberration, the *OCE* decreases together with the *EOD*, but more quickly for the large pixel than for the small one. Both curves attain a minimum, but the one for the larger pixel is achieved for a longer *EOD* value and it is deeper than for the small pixel. With an increasing amount of aberration, the *OCE* increases with the decreasing *EOD*. The interval of the *EOD* is about the same for both pixel sizes, except that graph is displaced by *0.03* to the longer *EOD* values for the larger pixel.

7.4. Defocus/Power

We present the *OCE* vs. *EOD* graph for the case of defocus aberration for two pixel sizes $2d = 3\lambda F/\#$ and $2d = 7\lambda F/\#$ in Figure 14. The first point on the left of each graph corresponds to the amount of defocus aberration $a_{20} = 1.25\lambda$, while the last point on the right of the graph corresponds to the case of no aberration.

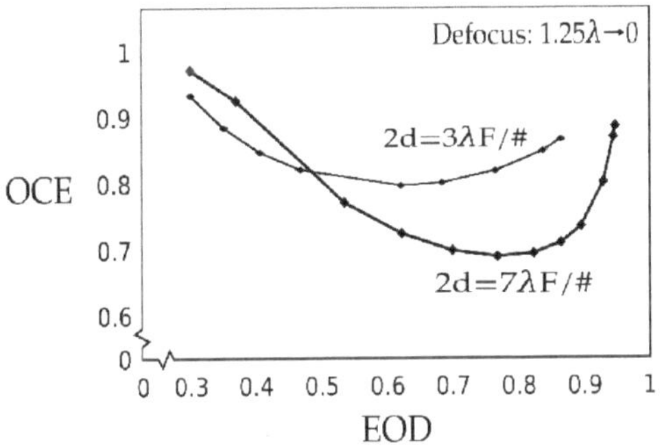

Figure 14. *OCE* vs. *EOD* graph for the case of defocus aberration for two pixel sizes $2d = 3\lambda F/\#$ and $2d = 7\lambda F/\#$. The first point on the left of each graph corresponds to the amount of defocus aberration, $a_{20} = 1.25\ \lambda$, while the last point on the graph corresponds to the case of no aberration, $a_{20} = 0$.

In the absence of aberration, the maximum *EOD* enclosed by the square pixel is *0.88* for the small pixel size $2d = 3\lambda F/\#$. In the absence of aberration, the maximum energy on the detector enclosed by the square pixel is *0.89*; for the case of the large pixel size, $2d = 7\lambda F/\#$. When the amount of defocus aberration increases to $a_{20} = 1.25\ \lambda$, the *EOD* decreases to *0.31*, when using a pixel size of $2d = 3\lambda F/\#$. When the amount of defocus aberration increases from 0 to 1.25λ, the *EOD* decreases *to 0.31*, when using a pixel size of $2d = 7\lambda F/\#$. The pixel size has no effect on the *EOD* for the large amount of defocus aberration.

When there is no aberration, the *OCE* achieves the value of *0.88* for the *EOD* value of *0.84* for the small pixel size $2d = 3\lambda F/\#$. When there is no aberration, the *OCE* achieves the value of *0.89* for the *EOD* values of *0.95*, for the case of the large pixel size $2d = 7\lambda F/\#$.

The *OCE* first smoothly decreases with increasing amount of aberration, achieving a minimum at *0.82* for *EOD* of *0.62* for the small pixel size $2d = 3\lambda F/\#$. The *OCE* rapidly decreases with increasing amount of aberration, achieving a minimum at *0.68* for the *EOD* of *0.77* in the case of the large pixel size $2d = 7\lambda F/\#$.

From the minimum, the curve for the *OCE* looks like a quadratic function with unequal sides. When the aberration further increases, the highest value is achieved for the *OCE*, equal to *0.95* at the *EOD* of *0.31*, for the small pixel size $2d = 3\lambda F/\#$. From the minimum, the *OCE* curve increases with the increasing aberration. The highest *OCE* value achieved is equal to *0.99* for the *EOD* of *0.31*, for the case of the large pixel size $2d = 7\lambda F/\#$.

The shapes of two defocus curves for the small and large pixel are in general quite similar. For a small amount of aberration, the *OCE* decreases together with the *EOD*, but more quickly for the large pixel than for the small one. Both curves attain a minimum, but the one for the larger pixel is achieved for a longer *EOD* value, and it is deeper than for the small pixel. With the increasing amount of aberration, the *OCE* increases with the decreasing *EOD*. The *OCE* for the large pixel starts to climb up slowly to the highest possible value of 1, but it cuts off at *0.99* for the *OCE* final value. For the smaller pixel, the curve is still increasing sharply at the maximum aberration, achieving *OCE* = *0.94*, while it slowly approaches to this value. The interval of the *EOD* is about the same for both pixel sizes, except that graph is displaced by 0.03 to the longer *EOD* values for the larger pixel.

8. Summary and Conclusions

In the cases of the individual aberration that we evaluated, we note that the effect of aberration is less pronounced on the *OCE* than on the *EOD*. Most importantly, the conventional wisdom would expect that the *OCE* increases with the *EOD* linearly regardless of the pixel size and the type of aberrations. In this work, we demonstrate for all aberrations that the *OCE* increases with the *EOD* for small amounts of aberration, except for the smaller pixel size in astigmatism, in which the asymmetrical nature of *psf* cannot be collected completely. However, as the amount of aberration reaches a certain point (as the *EOD* keeps getting smaller), the trend reverses and the *OCE* starts to increase. Furthermore, when the pixel is sufficiently large to include enough energy, this correlation is even more pronounced.

With respect to the pixel size, the most interesting aspect of the comparison of the *OCE* vs. the *EOD* graphs for two pixels under study is that they are quite similar; however, they appear to be compressed and slightly displaced along the *EOD*-axis.

The *EOD* is more sensitive to the amount of aberration than the *OCE*. The aberrations and their amount affect the *EOD* differently for both pixel sizes studied here. For smaller pixels, the aberration amount ranging from 0 to 1.25λ subtends quite a large interval on the *EOD* axis, from *0.25* to *0.95*, while the *OCE* domain varies from *0.80* to *1*. The width of the *EOD* interval is approximately the same for all aberrations studied here. The same amount of aberration, that is, $a_{ii} = 1.25\lambda$, decreases the *EOD* for the spherical to *0.24*, the coma to *0.29*, the defocus to *0.30*, and the astigmatism to *0.34*. Therefore, if the *EOD* is to be maximized while allowing some aberration, the astigmatism is the aberration to permit when imaging with a smaller pixel.

For the larger pixel, the aberrations cover quite a large interval on the *EOD* axis, while the *OCE* varies from *0.68* to *0.95*. The width of the *EOD* interval is significantly different for all aberrations studied here. The same amount of aberration, that is, $a_{ii} = 1.25\lambda$, decreases the *EOD* for the defocus and the astigmatism to *0.31*, for the spherical to *0.40*, and the coma to *0.60*. Therefore, if the *EOD* is to be maximized while allowing some aberration, the coma is the aberration to permit when imaging with the large pixel.

Based on these studies, we consider that the energy enclosed within the square pixel, the *EOD*, remains an interesting parameter but not necessarily the most informative one as an aid in design specifications. However, due to the high probability that the pixel center and the axis of the optical system are misaligned—for all pixels in the focal plane array—the energy that spills on the neighboring pixel is usually not lost to the signal. It is added to its nearest neighbors, where it most likely still contributes to the signal. This is particularly relevant for those cases where the small pixel intercepts small amounts of energy due to a large amount of aberration producing a large signal spread.

As a design guideline, then, we see that a small pixel $2d = 3\lambda F/\#$ results in a high *OCE* even for a large amount of aberration for all aberrations studied. For zero and very small amounts of aberration, the large pixel $2d = 7\lambda F/\#$ in general achieves a better *OCE* than the small pixel. In the intermediate aberration region, there does not appear to be much difference in the *OCE* for the two pixels studied here, except in the case of the spherical aberration, where a small pixel is preferable for most of the aberration interval.

In conclusion, for pixels larger than $2d = 3\lambda F/\#$, a small pixel will feature better performance when expecting jitter, misalignment, and other environmental and unpredictable external conditions for all but minimal aberration. When counting on the perfect aberration-free design, the choice of lager pixel is expected to have better energy collection properties, but such design will be plagued by the deterioration of another figure-of-merit, the image resolution.

In the future, we intend to extend this study to a mixture of aberrations to mimic a realistic optical system as close as possible. In a follow-up publication, we will address a mixture of aberrations on the performance of an instrument in an environment subjected to jitter and misalignments. With that, the same amounts of aberrations will be applied to both refractive and reflective optical systems so that we can explore and have better insight over the *OCE* vs. *EOD* relationships. We also plan to apply this theory to other remote sensing instruments.

Author Contributions: All authors contributed equally to this MS. Conceptualization, Y.W. and R.M.; Investigation, Y.W.; Methodology, M.S.; Visualization, M.S.; Writing—original draft, M.S. and Y.W.; Writing—review and editing, M.S.; Writing—final, M.S. All authors have read and agreed to the published version of the manuscript.

Funding: This research received no external funding.

Institutional Review Board Statement: The study did not require ethical approval or institutional review.

Informed Consent Statement: The study did not require informed consent.

Data Availability Statement: The data presented in this study are included in this study.

Acknowledgments: MS thanks RMA and BBM for helpful discussions.

Conflicts of Interest: The authors declare no conflicts of interest.

References

1. Strojnik, M.; Bravo-Medina, B.; Martin, R.; Wang, Y. Ensquared energy and optical centroid efficiency in optical sensors: Part 1, Theory. *Photonics* **2023**, *10*, 254. [CrossRef]
2. Muñoz, J.; Strojnik, M.; Paez, G. Phase recovery from a single undersampled interferogram. *Appl. Opt.* **2003**, *42*, 6846–6855. [CrossRef] [PubMed]
3. Brady, D.J.; Gehm, M.E.; Stack, R.A.; Marks, D.L.; Kittle, D.S.; Golish, D.R.; Vera, E.M.; Feller, S.D. Multiscale gigapixel photography. *Nature* **2012**, *486*, 386–389. [CrossRef] [PubMed]
4. Sroubek, F.; Kamenicky, J.; Lu, Y.M. Decomposition of Space-Variant Blur in Image Deconvolution. *IEEE Signal Process. Lett.* **2016**, *23*, 346–350. [CrossRef]
5. Catrysse, P.B.; Wandell, A.B. Optical efficiency of image sensor pixels. *J. Opt. Soc. Am. A* **2002**, *19*, 1610–1620. [CrossRef] [PubMed]
6. Li, X.; Suo, J.; Zhang, W.; Yuan, X.; Dai, Q. Universal and flexible optical aberration correction using deep-prior based deconvolution. In Proceedings of the 2021 IEEE/CVF International Conference on Computer Vision (ICCV), Montreal, BC, Canada, 11–17 October 2021; pp. 2593–2601.
7. Kerepecký, T.; Šroubek, F. D3Net: Joint demosaicking, deblurring, and deringing. In Proceedings of the 2020 25th International Conference on Pattern Recognition (ICPR), Milan, Italy, 10–15 January 2021; pp. 1–8.
8. Welford, W.T. *Aberrations of the Symmetrical Optical System*; Academic Press: London, UK, 1974; p. 120.
9. Born, M.; Wolf, E. *Principles of Optics*, 7th ed.; Cambridge University Press: Cambridge, UK, 1999.
10. Mahajan, V. *Aberration Theory Made Simple*; SPIE: Bellingham, WA, USA, 1991; Volume TT6, p. 11.
11. Smith, W. *Modern Optical Engineering*, 4th ed.; McGraw Hill: New York, NY, USA, 2008; p. 311.

www.ingramcontent.com/pod-product-compliance
Lightning Source LLC
Chambersburg PA
CBHW041343210926
51850CB00013B/2964
* 9 7 8 3 7 2 5 8 4 5 3 6 1 *